Table of Contents

The Logarithmic Unfolding of Astronomical Time: From Quantum Reality to Life

Michael A. Susko

AllrOneofUs Publishing
Baltimore, MD & Huntsville, A

The Logarithmic Unfolding of Astronomical Time: From Quantum Reality to Life

First edition, December 30, 2024.

To budding astronomers and cosmologists around the world.

Preface

In grade school I experienced a strong intimation that I would one day write a story of the universe, starting with its stars and planets. Perhaps this dream was planted because my father was a NASA aerospace engineer, and we lived in Huntsville, Alabama—called the "Space Center." It's a vision I've only come to realize now in my retired life.

My academic interest in chronology began decades ago with history, using generations of fifteen-year periods to map events and discover its phases. Over time, this evolved into a fascination with the history of life, measured in vaster time scales of millions and billions of years. Mapping vast timescales led me to use logarithms—a mathematical tool useful for understanding patterns in time.

While not often reflected upon, logarithms are a surprisingly simple concept. We see a tree outside that starts with a trunk, becomes several branches, then lesser branches, twigs and leaves—logarithmically expanding each time in number. The same pattern appears in our arms and hands: the single upper arm bone becomes two at the elbow, eight in the wrist, and fourteen in the fingers. Hurricanes and decibels of sound are mapped logarithmically—we could go on.

Now, we will use logarithms to map the more abstract phenomenon of time. Using a base of two, we can scale numbers by doubling or halving values—creating nodes or markers for time. In biological evolution, we started with the origin of life at 4 billion years ago, then kept halving to 2 Gya, 1 Gya, 500 million years, and so on. Remarkably, evidence shows major change clustering around these halving points, revealing a pattern of accelerating change.

This realization made me wonder: could a similar mapping extend further back in time to encompass the evolution of the universe? Could nodes at 16 Gya, 8 Gya, and 4 Gya mark qualitative leaps in cosmic complexity? While earlier evidence seemed insufficient to me to confirm this, the advent of the James Webb Space Telescope and other advances opened a door—showing that logarithmic patterning could be extended further back into astronomical time.

In short, this work explores how halving "Major Nodes" aligns with the universe's great transitions—from its beginning, to galaxies, to planets, to life—inviting us to see the cosmos through a new lens of rhythm and order. Using more scientific language, this work proposes that key transitions in cosmic and biological history cluster near base-2 halving points (...16, 8, 4, 2 Gyr...), a *hypothesis* we will explore against observations.

This charting does not explain how such events happened or the dynamics behind them. Rather, its task is more humble: simply to order events in logarithmic fashion on a time chart and see what they reveal. This may be termed a "descriptive heuristic"—a practical way to organize evidence to make things clear. Thus, we focus on the *pattern* revealed by the evidence, which, through the work of others, may help generate hypotheses about how and why.

The advantages of logarithmic, or log, charting are many. The first major one is that we can readily grasp the Major Nodes of change around which qualitative leaps occurred. Once focused on such nodes, we can then organize the detailed surrounding data, and avoid being left with mental confusion. In this work we will focus on five Major Nodes, aligning evidence which falls before and after.

Second, we are now able to easily view a wide range of data radiating from these central nodes. Further, we can readily sense their mathematical relationship, which gives order and meaningful sequence to the data. There is an important, perhaps even hidden, reality to keep in mind as you view a logarithmic chart. If we find events are relatively evenly spaced, the log chart is showing us that events are regularly *accelerating* in time. This is because each interval on a logarithmic chart represents a progressively smaller unit of time. Thus, the further we go to the present, we find more events happening in shorter frames of time. This proves visually and mentally helpful, as data doesn't crunch up toward the end of a chart and become hard to interpret.

A third advantage of log charting is that meaningful clustering around a given node means that every event helps to inform the others as to their place and meaning. When a gap in the evidence is revealed on a logarithmic chart, we can readily hypothesize what might be there. When new evidence presents itself, we can immediately see its relationship to what happened before and after. And once we have established our focal nodes and calculated how far and on which side of the node we are, we can make predictions. An important value of this model—not to be underestimated—is that it is predictive.

Generally, the movement is from uncertainty to certainty as to the qualitative change hypothesized at the Major Node, which itself falls in the logarithmic center. Critically, we provide logarithmic mapping to either side. We will map three precursor Minor Nodes, the Major Node in the middle, then three explicit Minor Nodes on either side, at 1/10, ¼ and ½ log distance.

This brings us to a fourth advantage. The chart not only shows leaps with the Major Nodes, but a gradual buildup and the aftermath of confirmation in logarithmic fashion. Of course, this is only apparent gradualness, for as previously noted, a log scale actually reflects acceleration. Thus, logarithmic charting compresses time intervals as we move forward, representing an increasing density of change. Thus, this ability to illustrate both leaps of change and the apparent "gradualness" of change can revolutionize our modeling and understanding of time. We will soon see for ourselves how this works, and it's best shown by illustration.

A final advantage is that log charting, in providing a clear framing and ordering evidence, enables us to tell a story. We should not underestimate the ability to create a coherent, accurately sequenced story, a narrative which suggests causation and provokes hypotheses.

Modeling by a logarithmic chart, with precursor and explicit nodes, reveals this regular unfolding. Narrative becomes a powerful tool, emphasizing the main line of action while leaving space for intricate developments.

In summary, a log chart offers a novel way to model time, with its ability to show leaps and how these changes are paced, reflecting a coherent and orderly manner. By also allowing evidence to inform itself, fill gaps, and frame the story around clear nodes, it invites us to interpret the past, and make predictions for the future.

In doing this work, I researched the literature extensively and had conversations with AI, which provided factual clarifications and support for inferences. Consider these two paragraphs, which discuss how we might present this model to young people.

> Framing cosmic history through the *logarithmic model* offers a wonderfully intuitive and coherent narrative for young minds and those new to these concepts. It's a framework that emphasizes observable milestones and patterns of complexity, making the universe feel accessible, ordered, and exciting, rather than overwhelming or overly speculative.

In sum, we can articulate our mission as twofold. 1) Provide a new model having a clear factual basis for understanding the history of the early universe. 2) Show the model's usefulness in revealing patterns clearly, generating hypotheses, making predictions, as well as leading to an excitement for discovery.

It is up to you to evaluate whether the evidence sufficiently supports a logarithmic understanding of our unfolding universe, and whether it grants you insight. It's also up to you to consider what such patterning might mean, and what it might lead you to imagine further.

Introduction

To some extent the astronomical world—paired with intricate and surprising properties of matter and energy fields—appears more distant and harder to understand for us who are biological beings. We understand reality and time from the perspective of embodied beings. It may be that we even inhabit a different kind of time than the astronomical world. Nonetheless, the world before life can be seen as having the capacity for life within it—so that in broad terms, some measure of harmony must exist, allowing us to approach this reality.

The scope of time is vast. The universe has existed for at least 13.8 billion years, and recent findings by the James Webb Space Telescope suggest it may be even older. This is an immense amount of astronomical time, and it's hard for us to comprehend. It gives us pause to consider how we might understand it and the evolutionary sequence that unfolds. I suggest that a logarithmic chart will provide us with a ready handle on the flow of time and will show that events tend to cluster around consistently spaced nodes of time.

To do this, we will map events as accurately as possible from the proposed start of the universe to the appearance of life. Our tool is logarithmic mapping: we simply halve values—using decreasing exponents or powers of two—to chart key transitions. We find that a relatively simple formula greatly aids in ordering and understanding events across these vast spans of time.

It may be surprising to find a mathematical regularity in time. Perhaps it should not be, for we expect to find space ordered. The unity of space-time, as presented in contemporary physics, suggests such a continuum. That is, if we can describe space as ordered, perhaps time is ordered as well.

To evaluate this hypothesis, we need to determine as precisely as possible the dates for events and place them on a logarithmic scale. After collecting as much relevant data as we can, we then consider how the evidence sorts out. As already described, no elaborate theorizing as to the cause of the phenomenon is necessary. Rather, our hypothesis is about patterning, seeing how the evidence fits and whether future evidence conforms or not.

The first stage of any theory of how and why things occurred is to consider direct, observable or empirical evidence and determine what patterns are present. With this knowledge in hand, you can more profitably theorize whatever model you propose.

It is a grounding step to return to the evidence. This may prove useful when old theories keep encountering new evidence, which creates tensions—leading to ad hoc additional components and parameters to make things work. Like in the scriptural saying, it is better to put new wine into new wineskins rather than patch up the old, lest the old one burst. Maybe we don't know what new wine to use yet, but we propose that our log charting provides a new skin, from which a more coherent theory can emerge.

However this skin is not totally new, as cosmologists have already identified a basic logarithmic construction to the ordering of the universe. The crucial formula that links the redshift in light from distant galaxies to the scale of the universe shows as much. This simple formula, $z=(1/a)-1$, hides a basic logarithmic reality: z is the ratio between observed and emitted wavelengths of light, and a is the scale factor, a measure of the universe's size.

Basically, this means that—if we frame things as powers of two—every time the wavelength of light doubles, the universe halves in size. Conversely, when the wavelength halves, the universe doubles in size. This combination of halving with doubling makes a strong case for a logarithmic reality at work in the universe's structure. Multiple independent data sets confirm this relationship. Because the size or scale factor relates directly to how far light has traveled, with a model of expansion we can determine the universe's rate of growth and its age.

This means our logarithmic numbering may be more than a charting device—it may be built into the fabric of the universe.

Having set out a rationale for a logarithmic approach, we turn to the practical steps for constructing and applying our logarithmic chart based on powers of two. As already intimated, we start by halving values to identify our Major Nodes in time. We will consider five Major Nodes in all: 32, 16, 8, 4, and 2 billion years. These nodes are easy to identify as we simply halve values.

However, we will need to find out exactly how far a given piece of evidence falls from these nodes. To do this, we use powers of two and a calculator. For example, to find the Major Nodes, you set the base as two, then use the **y^x** button, which raises the base to the desired power. For instance, 2 to the 33rd power brings you close to the 4 billion marker. I ended up using $2^{32.9}$ to land closer to 4 Gya.

From there, you need to determine how far an event is from the given node. We have chosen to highlight *Minor Nodes*—set, shorter distances on either side of the Major Node—mainly 1/10, ¼, and 1/3 log distance on either side. You can calculate this by either subtracting or adding 0.1, 0.25, or .33 to the exponent in the power of two, then using these numbers to construct a logarithmic chart. In sum, creating Minor Nodes allows us to place evidence and see how it aligns in proximity with the Major Nodes.

What is often challenging is finding the most precise dates possible for events, which in astronomical dating can include a large margin of error. Once this is done, we can see if the evidence collects or points to certain nodes in time. The chart will also reveal whether the evidence shows a major shift at certain logarithmic intervals. This begins a more interpretive phase, but one grounded in observable evidence.

As mentioned earlier, this logarithmic model was developed from the exploration of patterns found in biological evolution. Let us consider four of them:

I. Major Nodes of Change

There are leaps of change in greater complexity, following a logarithmic pattern of regular pulses, which we refer to as "Major Nodes." These nodes occur at regular intervals, which are halving in time as we come to the present. This is the pattern we project for discrete qualitative leaps, one which is accelerating in time.

II. Logarithmic Pattern of Evidence

The logarithmic model incorporates certainty, uncertainty and lack of evidence. That is, we expect events to have "explicit" evidence increasingly after the proposed node of change. Prior to that, we have precursor events, suggestions of change building, which move from uncertainty to certainty. This evidence patterns logarithmically.

A logarithmic chart may appear misleading. We see the same linear distance on the chart, such as with the ¼ log node on either side. But the same distance before the Major Node covers more time, while after the Major Node, the same 1/log node distance covers less time. This implies that if we find evidence regularly filling our chart, more is happening in less time. Put another way: a longer build-up *precursor* period before the Major Node has more indefinite events, while a shorter *explicit* period afterwards is confirming of events.

III. The Increasing Precision in Logarithmic Charting

Because the Major Nodes are regularly halving in temporal distance, and as our chart is becoming more refined in its dating with smaller and smaller units, we need to be more accurate as we move toward the present. Thus, a logarithmic model reveals not only the acceleration of events, but an increasing *precision* to the timing of unfolding events. Thus, a log chart has a type of built-in confirmation momentum—a tightening of pattern—which remarkably continues at the same pacing into biological and early cultural evolution.

IV. Sequence of Four Nodes

In prior study, a sequencing in nodes was identified, for which I now offer phrasing for astronomical evolution. Its full relevance remains to be explored further.

1) A quantum reality that ultimately birthed things—a speculative pre-Big Bang reality

2) A phase of change characterized by rapid growth and a scale order increase in complexity and size: Galactic evolution

3) A more hidden stage of stabilization, maturation and transition—habitable planetary formation—which sets the stage for:

4) Another quantum change: the start of life.

This fourth value becomes the first of a new series, which we continue with elaboration in this work, detailing the appearance of eukaryotic and macroscopic life.

To review, the core concepts are these:

- *The Major Nodes,* or halving nodes, with qualitative leaps in complexity.
- *Precursor Nodes*: events which approximate qualitative change, mapped as Minor Nodes, from 1/3 to ¼ to 1/10 log distance prior to the Major Node.
- *Explicit Nodes*: events after the Major Node and are confirming of change, mapped from 1/10, ¼, to 1/3 log

distance.

Using the above, we can at a glance organize cosmological events with this charting. Let us first give a quick preview:

1. *32 Billion Years Ago:* Pre-Big Bang Period, a speculative node
2. *16 Billion Years Ago:* The observable universe, large-scale galaxies
3. *8 Billion Years Ago:* Habitable rocky planets
4. *4 Billion Years Ago: The Start of Life*
5. *2 Billion Years Ago:* Eukaryotic Life, Oxygenation of Earth

Now let us offer a more extended preview of the five Major Nodes, which we will use to mark astronomical time. These nodes, as previously described, result from a regular halving principle, obtained by using powers of two.

First Node: Pre-Big Bang and possible earlier Big Bang (~32 Gya, $2^{34.9}$)—Speculative placeholder.

This node involves considering a full node back in time before the "Big Bang," where we logically posit the existence of a prior matrix. A possible earlier date for the "Big Bang" could also be proposed, recently suggested to account for the unexpected finding of fully formed galaxies soon after the "Big Bang" event.[1]

Second Node: (16 Gya) Galactic Formation, Large-scale Structure Formation

The second node is confirmed by finding the first observable objects in the universe at 13.5 Gya, approximately 25% log distance after the node. Already, we find massive, luminous, and highly structured galaxies, with unexpected evolution of complex elements. Thick rotating disks have evolved by this time, along with the emergence of larger-scale galactic structures. Last, we find evidence of galaxies clustering and merging.

Mid-node: (11.3 Gya) Evidence for gaseous planets and simple rocky planets.

Between the second and third nodes, a transition stage can be identified with an early phase of planet formation, which we could term simple planets.

Third Node: (8 Gya) Layered, Interactive Rocky-surfaced Planets in the Habitable Zone.

At this node, rocky-surfaced, layered planets with interactive processes appeared. Importantly, enough are being produced to create a statistical argument that many exist within the habitable zone. Several lines of evidence coalesce that support this.

Fourth Node (4 Gya). The Emergence of Life

The fourth node marks the emergence of a new order of complexity, as living matter appears. This event represents a clear quantum leap at a precisely demarcated time.

Fifth Node (2 Gya). The Eukaryotization of Life

In proximity to 2 billion years ago, we posit a heightened complexity. Specifically, three interlinked events are explored: the oxygenation of the atmosphere and oceans, the activation of global plate tectonics, and the emergence of eukaryotic life. The latter creating a new interactive multiplicity within life, making our final log order of increasing complexity in physical-living systems.

Interestingly, we can frame each of these five nodes as marking a new level of complexity in matter. Thus, in the broadest sense, we offer this picture of emerging complexity.

1) *Simple Matter*, a speculative type of matter existing before the Big Bang, the matrix that produced hydrogen and helium.

2) *Enriched Matter* present in galaxies at the start of the observable universe, which has more complex matter or elements, like carbon, oxygen, and iron. It's hypothesized that these were synthesized shortly before in a period of star formation.

3) *Dynamically Enriched Matter* can be used to describe complex organic molecules that held pre-biotic potential. These molecules could react with a wide variety of other molecules, forming stable, versatile bonds that held the potential to build more complexity. In the case of carbon, its four covalent bonds make for versatility and amazing connectivity.

4) *Living Matter* exhibits a clear quantum leap in complexity. Briefly, life is a system that has the capacity to intake substances, metabolize them to create its form and energy—and to maintain this markedly more complex structure in a simpler environment.

5) *Interactive Living Matter* represents a major leap in complexity, which led to macroscopic forms of life. At this stage life forms merged with other life forms to produce a novel integration, a symbiosis that qualitatively transformed life's capabilities—as if a new state of life had been achieved.

In sum, we will find that by anchoring our finds around these nodes, a surprising clarity emerges. They will help us to see clearly and understand the story of the universe and its events so distant from us in space and time.

With this basic understanding of our model in place, we are now ready to traverse our cosmological journey, to visit the Major Nodes in time and see how the universe unfolded.

CHAPTER I

1st Node (32 Gya)
Earlier and/or Pre-Big Bang Universe

The current consensus model starts the universe at 13.8 Gya with two pulses, a rapid inflationary phase of space-time and then a hot Big Bang.[2] But our modeling seeks to open the door to consider if events such as these happened earlier, and what happened before the observed universe's actual start.

In the past several years, evidence has arisen that the Standard Model did not predict, and demanded the theory be altered in ad hoc ways. For example, to make sense of certain observations, large amounts of dark matter and dark energy have been theorized. Midway in the universe's history, dark energy has been employed to account for an accelerating expansion of the universe. These two pillars of modern cosmology, however, still lack direct particle or field detection after decades of searching. This doesn't mean that these theories haven't provided a useful frame that fits an array of evidence. But the lack of direct evidence should give us pause and allow us to roam more freely. [3]

Discrepancies in theories, modeling and actual observation in two other areas also suggest that the verdict is still out as to the universe's start time. An important discrepancy, called the Hubble Tension, is related to how fast the universe is expanding, which helps determine its age-Local measures of the space stretching between two galaxies do not match the global measures for expansion, with an 8 to 10% difference. These measurements are taken by measuring the amount of stretching in the light measured, called the redshift. Thus, the primary evidence for the universe's age and expansion is the redshift we see in light we observe from the distant universe, where those frequencies increase the further the sources are away in time and space.

In another major discrepancy, the amount of energy actually found in quantum space is orders of magnitude less than quantum modeling predicts. This "cosmological constant problem" has been put on a set-aside shelf and tolerated as not directly impacting the rest of cosmological theory, but it suggests that at some basic level, all the story is not in.

Current cosmological theory may eventually be updated to a broader-based theory, for which several possibilities have emerged. It is in a spirit of exploration that I propose a speculative node at 32 Gya—a conceptual placeholder as to what preceded the Big Bang, while also providing room for an earlier revised start date for the universe.

Recently, the "startling" discovery of early bright, massive galaxies as early as 280 million years after the "Big Bang" has created what Gupta calls the "impossible early galaxy problem."[4] Galaxies in canonical models normally take 1 to 2 billion years to form. The observed early formation of early galaxies suggests the standard model is at risk of being able to account for basic phenomena. When a theory is "startled" by new data, or has a whole swathe of findings not predicted by their theory, then it's time to reevaluate. And galaxies are no small feature of the universe, being its basic building block and holder of the universe's matter and energy.

For the purposes of exploring a potential logarithmic patterning, we will propose the possibility of a new node extending much further back in time. A proposed start date has recently been offered by Gupta at 26.7 Gya, almost double the conventional date. This was, as just intimated, prompted by the finding of early galaxies. However, with a 32-Gyr Node, we are projecting back in time a period for which we have no observable evidence.

Thus, this node is a placeholder, for if not an earlier stage of the universe, something had to be here. If we accept the philosophical and scientific approach that something can't come from nothing, we can safely say something had to be there. But we can do more than this; we can, I believe, safely project a very basic existing pattern back into time.

Thus, relying on patterning observed from biological evolution and the pattern going forward in astronomical evolution, we project an earlier node logarithmically. If we started with the advent of life at 4 Gya, we can, by simple doubling, next go back to 8 Gya, then 16 Gya, and finally 32 Gya. An early phase of galactic maturity appears within 25% log distance of the 16 billion marker, indicating that node is secure. We know with certainty that life is proximal to 4 Gya. For the 8 Gya, we hypothesize major planetary change, and we will see what evidence emerges.

What about the period before the Big Bang, the first proposed node? There had to have been a *before,* and we use the logarithmic marker, 32 Gya, to indicate this before period. In a sense this node is the most invisible of all, reflecting the uncertainty we previously mentioned in our four phases of Major Nodes.

Importantly, Gupta has offered an updated theoretical model that fits with an earlier node. His proposed date of 26.7 Gya falls at 25% log distance from the projected 32 Gya node. With the ¼ explicit Minor Node after the Major Node, we expect evidentiary confirmation. For this atypical node——the start of everything, and before anything observable — we can only infer from preexisting patterns and offer plausible theoretical forms of confirmation.

This, first of all the nodes in our logarithmic study, is unique in an evolving universe. We detail further two observations at this point which relate to the hyper-invisibility of this proposed node.

1) An expectation of invisibility: We expect a degree of invisibility prior to any major evolutionary event. That is why our model typically allows for evidence to emerge within 25% log distance after the "Major Node." We also expect more invisibility with the start of any four

series of nodes, as they represent a more quantum change. Then, the following nodes offer increasing clarity. Last, with the start of all things and the beginning of the universe as we know it, we expect this invisibility to be further heightened.

2) A node can reasonably be inferred. Even though it is observationally invisible and remains conjectural, we know something must have existed during this time frame. To repeat, from all our experience in this universe, something does not come from nothing. Thus, there had to be a before, and this node can be used to represent that.

Before we conjecture about this earlier time, let us review in more detail what our first evidence shows us—bright, massive galaxies, and many of them, at 13.5 Gya. So already within 25% of the earlier 16 Gya Node, we have complex systems in place. Such galaxies, as we have mentioned, typically require at least 1 to 2 billion years to form, with mature galaxies taking 6 to 8 billion years to reach full development.

This presents a problem for the standard start date of 13.8 Gya. Recent observations suggest that events are occurring much faster than current models predict—showing mature-like galaxies just 300 million years after the Big Bang, compared to the expected timeline of 6 billion years.[5] Even under accelerated formation scenarios for young galaxies, we are conjuring a pace that is 2 to 3 times faster than what current galactic formation theories can explain. Aside from time scale, there is also a mechanics issue. These ultra-massive galaxies require two to three times the norm of baryonic matter — which must be converted into stars, two to three times the norm.[6] To repeat, the excess of these ultra-massive galaxies in terms of size, luminosity, and number is contrary to our expectation of the standard model.[7] All in all, significant revisions are in order regarding our theory of galaxy formation if we are going to keep the current start date of the universe.

One more technical factor calls things into question. An important measure in galaxy processes is star formation efficiency (SFE), the percentage of gas converted into stars. Current-day galaxies convert about 1–3% of their gas, while early systems, to form massive galaxies as observed, would require a star formation efficiency several to ten times higher than normal. This logarithmic order of difference suggests our current cosmological model is incomplete or requires significant revision.

This finding opens the door to considering an earlier start date for the universe and new models. Gupta has offered a theoretical model to account for the new data, which falls within our 32 Gya frame. His CCC+LC model combines the universe expanding to make for a red-shift as the standard model uses, and a "tired light" hypothesis. The latter means some of the redshift is due to light passing through the universe, interacting with particles and energy through the vast spaces—the effect of its energies being diminished, helping to create a redshift.

"Tired light" may not be the best word for this, as its negative connotations may bias us against the theory. Perhaps it's better termed "Light Energy Exchange"—what happens when light and matter interact over vast distances. We know light becomes "tired" over shorter scales in varied ways. The question is whether it adds up to be significant on the cosmological scale.

Some recent evidence suggests light is impacted more than has been allowed. Redshift is the primary indicator of distance in the universe, and recent evidence has surfaced that the rotational velocity of a galaxy affects redshift. That is, redshift measurements vary consistently depending if the galaxy is rotating the same way as our galaxy or in the opposite way. This means that light can be influenced as it journeys past galaxies in support of the "tired light" or "Light Energy Exchange" hypothesis.[8]

We add two more points. Gupta's model has the big advantage of not having to rely on dark matter or dark energy, which are added as unobservable entities to explain the universe's beginning.[9] The model's biggest drawback currently is that it doesn't explain the Cosmic Microwave Background or CMB. Perhaps we should not expect all problems to be resolved in any given model, and that such models will only give us partial understanding before the complexity of the universe. For now, we simply say that we should begin to seriously consider hypotheses that propose earlier start dates for the universe.

So aside from truly alternative modeling or variations of the Big Bang, which sometimes push back toward the 32 Gya Node, what more can we say? That is, can we describe the period before any "Big Bang" or start of the Universe, which we will by definition have no direct observable evidence? I offer three initial traits, of which the first has already been broached.

1) Something existed before the observable universe, unless we say something comes from nothing. That is neither philosophically nor scientifically sound.

2) There is pattern in the pre-universe. At all levels and periods of time, we see pattern in the universe. There may be periods of relative chaos, but they still exhibit considerable pattern. Often, too, relative chaos can have an emergent creative result in a higher order of complexity. Thus, we reasonably hypothesize pattern would be present at the proposed node of 32 Gya, although the full nature of that pattern is not clear.

3) In the pre-universe, power and/or the potential for power was present. The universe as we know it and at all observable times, has immense energy present. Such energy was expressed at the time of the Big Bang. And as the universe expands, the total energy in the empty space between galaxies increases. Thus, energy, or at least the potential for immense energy, had to be present before the Big Bang.

In sum, we posit at least three descriptors for the pre-"Big Bang" world: 1) Presence, 2) Pattern and 3) the Potential to generate immense energy.

To this we can now add a fourth element: *a logarithmic capacity.* We posit that each of these elements had built-in capacities for logarithmic expansion or expression. We argue that a fourth trait overarches the other three. Let's now consider the logarithmic quality of each of the traits, going from the third to the first:

1) A logarithmic expression of power is inherent in the universe as it gives rise to exponential expansions of energy and matter. The current model hypothesizes an initial exponential spatial inflation to the universe and continued cosmic expansion due to dark energy.

2) We posit a logarithmic quality to time, that things unfold with regular leaps in complexity, which this work will illustrate. Thus, the prior node either is governed logarithmically the potential to generate logarithmic time.

3) Presence. There can be many aspects to presence, but we take one of these to be scale. There are logarithmic scale distances on the cosmic level between star systems, galaxies, clusters of galaxies and superclusters. There is also a logarithmic scaling that paradoxically puts log increases of complexity into smaller logarithmically decreasing spaces, as when we go from stars to planets and then from planets to life.

All in all, we posit a logarithmic nature to the structure of the universe. Scientists often use logarithms as a tool in their equations to discern things. We propose that this tool actually reveals a deep structural trait to the universe. Thus, we also hypothesize that logarithmic reality extended back to the earlier proposed node.

In sum, the first "proposed" node at 32 Gya serves as a placeholder for the reality that existed before the Big Bang. We can begin to discern the outlines of that reality, but much remains to be explored. We are ready now to move to the observable universe.

CHAPTER II

2nd Node (16 Gya)
Galactic Emergence & Formation

The 10% Node: Early Signals

Our logarithmic model suggests that within 10% of the 16-Gyr node, we expect a signal marking the universe's true beginning—not absolute confirmation, but highly suggestive. The best possibility for such a signal is the dating of globular clusters, dense groupings of stars ranging from hundreds of thousands to a few million. The oldest stars in these clusters show extremely low metallicity, indicating that more complex elements had not yet had time to evolve.

So how old can they be? This is not a trivial question. Globular clusters give us an opportunity to fact-check the actual age of the universe, as these early formations can be independently dated. Let me repeat this point by citing a recent article titled *The Absolute Age of Milky Way Globular Clusters:*

> Globular clusters are the oldest objects whose ages may be accurately determined and can set the lower limit of the age of the universe independent of the cosmological model.[10]

For decades, from the 1970s to the 1990s, multiple independent teams, using different stellar models and photometric methods, consistently placed the oldest globular clusters between 14 and 19 Gya with 16.3 Gya as the mean. This was widely accepted and considered a hard boundary condition––a finding that any theory for the universe had to account. For example, M92 age was dated at 15 Gya using a "long" distance scale and higher values for a shorter scale.[11] This finding by Vandenberg reflects his focus on determining "absolute cluster ages" —that is, dating not depending on any theory or model.[12] In other words, globular clusters act as a fact-based anchor—offering independent, non-redshift-based evidence for the universe's age and one which we believe provides an early signal pointing to the 16 Gyr node.

However, since 2000 there has been a tendency to lower these age estimates, not due to any major error in the broader calculations. Revisions have aligned models to be in correspondence with the cosmological age, which holds that the universe can be no older than 13.8 Gya. Thus, studies use distance parameters, which depend on the Standard Model. In addition, studies often imposed a maximum prior (e.g. 15 Gyr), which prevents the data from naturally exploring older solutions seen in earlier literature. One study, setting the boundary at 15 Gya, found Milky Way globular clusters having an age distribution with an "old" component peaking at ~13.4 ± 1.1 Gyr and a statistical tail that extends that butted against ~15 Gyr.[13] The latter date, we note, falls at our 10% Minor Node.

In short, operating under the assumptions of the standard model entangles the data and finds estimates that tend not to allow for a higher estimate. It turns out there is a physics issue related to the helium diffusion of stars that should be allowed for and would typically reduce the derived ages by ~0.7 Gyr, with some variance.[14] Even here, if we subtract that from 16 to 15.5 Gya, we are at 15.3 to 14.8 Gya--still in the range of our 10% Node. To whatever figure is ultimately determined, we must also add time for a globular cluster to fully form—generally 5-30 million years, which moves us yet closer to the 16-Gya Major Node.

In sum, we can assign a signal from data that globular clusters are pointing toward the Major Node of 16 billion years ago. It is not definitive proof or outright confirmation--the evidence is uncertain and shifting. Yet, this is what we expect at this early stage, and for which we expect certainty by the 25% Explicit Node.

Within 10% log distance, we can also look for models that indicate the plausibility of a 16 Gya Node as the start time for the universe. Several alternative models have suggested start dates of 15.4 or 15.3 Gya, falling within this crucial range—constituting a direct alignment with our logarithmic modeling. These serve as a type of soft signal, compared to prior indicators from observational data. A compelling example is "An Alternative Cosmological Model for an Expanding Universe" by George and Johansson.[15]

In a nutshell, this model involves pairing a linear scale of the universe growing in size over time with logarithmic increments of time. This recalls our earlier demonstration that the z-shift formula encodes a logarithmic rhythm into the fabric of spacetime. An important feature of this model, however, is that both time and space expand together. Just as space is expanding between galaxies in the Standard Model, this theory allows for time being stretched or expanded.

Notably, this model removes many objections to the standard theory, such as the significant disparity between the locally observed level of energy in space compared to the global estimate. The equations used in this model show the universe's expansion passing naturally from quantum to cosmic scales. This means that this model does not require hypothesized quantities of dark matter and dark energy. This new theory is still in development but is mathematically robust, positing a flat infinite universe, with a start date at 15.4 Gya, just 5% log distance from our Major Node. If this model stands up to future scrutiny, the principle of Ockham's razor would favor its simplicity and consistency with observations.

As we might gather from the above, a model that logarithmically scales time—that links logarithmic intervals with linear time—resonates with the charting offered here. Interestingly, George's alternative model also posits that logarithmic time is real, not just a mapping device that makes things easier for us to read. That is, the model's increments of log time represent real, expanding intervals in the physical universe. Put another way, equal increments of log time, as we come closer to the present and which compose less linear time, mean that time is actually expanding.

Our paper shows that log charting is an effective way to structure events, but it remains agnostic whether time itself expands. Still, it's interesting that mathematics and dynamics of an alternative model align with our logarithmic charting. Thus, we remain open to the hypothesis that a logarithmic temporal reality underlies cosmic evolution.

Our main purpose, however, is more narrow at this point. That is, such models incorporating dynamics can serve as a soft signal for the 16 Gya Node as the start date for the observable universe.

The 25% Node: (13.48 Gyr): Confirmation

In our logarithmic hypothesis for the structuring of spacetime, we hold that the threshold for visibility, for clear evidence, emerges by 25% log distance after the Major Node. To review the math: if 2 to the 33.9 power is approximately 16 Gya for the Major Node, then 25% log distance after (or 2 to the 33.65 power) equals 13.48 Gya. This is precisely the time we find a plethora of hard evidence for galaxies in the universe. This modeling allows for an earlier period of uncertainty or invisibility, as with the globular clusters and inferring projected start times of the mature galaxies we find at this Minor Node.

If we allow for real time elapsed, the Big Bang provides only 0.3 Gya time for galaxy formation, versus 2.5 Gya in our logarithmic model—with the start node at 16 Gya. Importantly, we shall see that this larger gap proposed by the logarithmic model aligns more closely with the time necessary for mature galaxy formation, which we detect in the early universe. Thus, our goal here is to simply map evidence of factual sightings and see if there is logarithmic patterning. In making this correlation, we don't offer a theory of how things happened, nor do we need to make large, unproven assumptions to make the theory work. We will simply map events as accurately as we can in "logarithmic space-time." Thus, our mapping is model-free. It seeks to reveal the pattern, the relationship or time ratio between events—a mapping that can hold its weight even if the model changes, and even if time value for the z shift reads differently.

We now turn to the evidence that is revolutionizing our understanding of the early universe. The James Webb Space Telescope has detected very early galaxies, which are the major structural units of the universe. Here are some of the oldest currently dated ones—figures which may soon be surpassed. Galaxy MoM-z14, discovered at z = 14.44, corresponding to about 13.52 Gyr in look-back time. JADES-GS-z14-0, at z=14.32, follows closely at 13.51 Gyr. This is followed quickly by JADES-GS-z14-1, dated 13 million years later, or at 13.497 Gyr. These three galaxies are not isolated findings—numerous candidates fall in this narrow time range, with a handful now spectroscopically confirmed. Thus, right as expected, we have strong confirmation within 25% distance of our log node for the universe's emergence.

Alongside very young systems, we already see surprisingly massive, luminous galaxies. At this early time, at least three lines of factual evidence show that galaxies have achieved a an early level of maturity: high luminosity, massive galactic size, and an unexpectedly large number of galaxies. Thus, we can infer that galaxies had been evolving for a considerable period by the time of our observation.

To these three factors we can also add a fourth. Evidence has surfaced that our own Milky Way possesses a thick disk that is rotating. Although still metal poor, this disk named "PanGu" is kinematically—its movement—and structurally a disk, not a spheroid or purely random/thick system.[16] This structure thus represents a disjunct and significant leap in complexification.

If the one-fourth Explicit Node presents convincing evidence that the universe has evolved to this level of complexity, we are naturally led to ask more precisely: how much further back in time can we infer actual beginnings? In other words, when did these galaxies and chemically enriched systems first come into existence? These characteristics—mass, brightness, abundance—strongly imply their formation long before they became visible to us. Most accepted models of galactic formation have a buildup period on the order of a billion to a billion and a half years. This points to a cosmic origin in the range of 14.8 to 15.3 Gyr.

This confluence of new data has created a crisis for the standard model, whose elaborate modeling only works with a start date for the universe at 13.8 Gyr—a figure arrived at before the James Webb Space Telescope findings. In this light, consider the following statement from a peer-reviewed article

> Conventional explanations for these puzzling data require extraordinary and difficult-to-justify assumptions about galaxy evolution.[17]

We are currently witnessing the field of astronomy rushing to determine whether galaxy formation and other processes can be more punctuated—or occur more rapidly. It's an interesting parallel to the field of biology, which followed a gradual Darwinian model and resisted evidence of punctuated change, but which is now being increasingly accepted.

However, we do not need to resort to modeling gymnastics to accommodate novel evidence. We can simply let the evidence fall where it does—and see what it reveals. In this case, we argue that logarithmic scaling reveals the pattern of change more clearly. To review, we see unequivocal evidence for an early stage of maturity in galaxies at ¼ log distance from the Major Node, one whose backstory is increasingly pointing to the 16-Gyr Node.

We turn from galactic parameters implying early start dates to evidence pointing to the one-fourth Explicit Node as a pulse of complexity. By z≈10 (~13.3 Gya), spectra already show oxygen and other "metals" in galaxy gas. The overall oxygen abundance is lower than today—typically a few tenths of the Sun's level—but its presence shows that the basic galactic cycle (fresh gas inflow → star formation → supernova enrichment → partial metal retention) was already underway. Because short-lived massive stars forge and eject oxygen on timescales of tens to a few hundred million years, these detections are best read as evidence of rapid early enrichment.[18]

By the one-fourth Explicit Node, evidence for an enriching universe is also shown by the mass–metallicity relation. In short, bigger, more massive galaxies tend to have more heavy elements than smaller ones—the same pattern we see today—at roughly three times lower levels on average. Again, this means the basic galactic cycle—fresh gas inflow, star formation, supernova enrichment, and partial metal retention—is already up and running and starting to settle into a regulated rhythm. In short, maturing galaxies are acting as integrated systems by z ≈ 10 (13.32 Gya), about 0.16 Gyr after the 1/4 Explicit Node at 13.48 Gya.[19]

One-Third Log Distance: 12.72 Gya Further Confirmation of Galactic Evolution

By 33% log distance (12.72 Gya), several lines of evidence indicate that galaxy formation has matured. We see increased morphological variety: rotation-supported disks—sometimes with spiral-like patterns—alongside the clumps seen earlier. Feedback is evident within galaxies: AGN (Active Galactic Nuclei), driven outflows, starburst winds, and metal enrichment is clearer. In short, many systems have become more organized, and some cases are already moving to reduced star formation.

A recent survey of 39 early galaxies sums up their findings from the period between 12.8 to 12.5 Gya (from 4 to 6 z redshift)—straddling either side of our 1/3 explicit node (12.72 Gya). The abstract states:

> The survey reveals diverse morphologies and kinematics, including rotating disks, merging systems, [Cii] emission tails from potential interactions, and clumpy star formation. [20]

Galaxies at this stage are not just static, homogeneous rotating disks, but they host increasingly complex chemical and physical processes. Their interstellar mediums have complexified—harboring processes such as shock-excited gas, outflows, and widespread neutral and ionized phases contributing to [Cii] cooling—the latter abetting star formation.

The Cosmic Grapes Galaxy at z=6.072, or 12.87 Gya, is an example very near the 1/3 Explicit Node (12.72 Gya). This well-studied system provides unambiguous evidence for galaxy rotation and greater structural and dynamic complexity, harboring at least 15 compact star-forming clumps.[21]

At this stage, a major structural dimension in the Milky Way emerged: the formation of its early thin disk. By 12.5 billion years ago (Gya)—about 35% log distance after the 16 Gya Major Node—the Galaxy's disk reached a technical definition: the vertical scale height of stars dropped below half the radial scale length. This marked a transition from a more spheroidal, rounded shape to a distinctly flatter disk structure, indicating a significant leap in galactic order and stability.[22] This turning point mattered greatly, because by the next

Major Node at 8Gya, this thin disk matured further and provided the stable groundwork needed for the formation of potentially habitable rocky planets

Thus, not only do we see mature galaxies, but we detect candidate quiescent galaxies already reaching the *end* of their evolution. These galaxies appear near the one-third Explicit Node—such as JADES-GS-z7-01, dated to 13.1 Gya (z = 7.3).[23] A quiescent galaxy has ceased star formation, implying that its stellar population and structural maturity were achieved much earlier.

How long is a galaxy's lifetime? Some early galaxies, we note, may have formed and moved to quiescence remarkably fast, within 0.6 to 0.8 billion years, which pushes against the standard model's start time.[24] More typically, two or three times that duration is required, given the known timescales for massive star formation, supernova enrichment, bulge and disk settling, and the cessation of cold gas inflow. Drawing on a study of post-starburst galaxies, we can extrapolate a timeline—illustrative, based on intermediate-redshift post-starburst statistics—that extends the time needed to 2.9 billion years. This includes an average star formation phase calculated to be 1.5 billion years, followed by an average quenching phase of 1.4 Gyr.[25] The first phase takes us back to 14.6 Gya, while the second lands directly on our projected start node. In other words, if we observe a galaxy to be quiescent at 13.1 Gya, and then add the formation-to-quiescence period of 2.9 Gyr, we can infer an origin close to the 16 Gya. Thus, early quiescent galaxies provide strong inferential support for a 16 Gya origin framework—provided the model that accompanies this charting allows for real gains in elapsed time.

To review, the proposed marker for a Major Pulse in the universe's formation is within the logarithmic vicinity of the 16 Gya Node. We see evidence emerging at 1/10 to ¼ to 1/3 log distance, increasingly adding confirmation. We could have placed this as our first pulse of the observable universe. We are now ready to move forward to the next Major Pulse, which presents a further amazing leap of complexity.

CHAPTER III

3rd Node (8 Gya)
Rocky Planets in the Habitable Zone

As ultimately, we are looking toward a goal which culminates in life, we will follow the movement of the universe that moved toward creating habitable planets. This, admittedly, is a bias, that this work focuses on tracing the path that led to us. This does not mean that other facets of the universe that did not move in that direction are not of interest and shouldn't be explored. It just says that as designated observers and explorers of the universe, we have a natural interest in where we came from and where we are going.

At the 8 Gya Node, we are at a midway point in the age of the observable universe. Our logarithmic scale lands here and predicts a leap in complexity. Events on either side receive a lot of press around the Big Bang, within 25% of the 16 Gya Node, as well as the formation of our solar system and the emergence of life, circa 4 Gya. Let's turn our attention to this in-between Node at 8 Gya and see if significance presents.

At $z = 1$ (~7.82 Gya) right after the Major Node, the universe had expanded to twice its previous size and had undergone significant structural maturation. This redshift aligns closely with our predicted 8 Gya Major Node, suggesting a meaningful transition in cosmic evolution. At this milestone, we hypothesize the universe has matured enough and created enough numbers to make a qualitative leap in complexity. It matches the opening of an observational window when visible and near-infra wavelengths come into focus.

The change that ultimately illustrates this node is the emergence of rocky-crusted planets in the habitable zone. From the point of view of life's emergence, this stage is highly significant. A dynamic geosphere, or a highly interactive rocky planet, gave rise to lithospheres, atmospheres, and hydrospheres—all of which served to become prebiotic, and thus harboring the potential for life.

Planets should not be underestimated as a qualitative leap in complexity. Stars have a measure of complexity, but their process is basically singular, in which a gravity inflow is balanced by a gaseous outflow in a type of dynamic equilibrium. Planets, the kind that we are considering here have multiple processes which are coupled and occurring across vastly different scales. They are notable for having distinct layers beneath and above its surface, which interact physically and chemically with each other. Perhaps we should not be surprised that complexifying events in planetary evolution are greatly accelerated in time compared to galactic and stellar events. Another important clue to planetary complexity is their sensitivity to initial conditions, in which the natal disk can create a surprising array of different planetary types. Last, the complexity is such that an array of questions remains unanswered about planets, including the way in which planets were generated. In sum, this node focuses our attention on complex rocky planets without which we would not exist.[26]

Let us first set the stage for development. At least three lines of evidence converge, confirming that by z ~ 1 about ~8 Gya, conditions were suitable for widespread rocky planet formation. First, the level of metallicity, or the presence of more complex, "heavier" elements, had reached sufficient levels to readily make rocky planets. Second, the classic thin disk was in place by this time, a thinner, flatter disk making for more stable stellar orbits, which allowed complex planets to form and evolve. Third, sufficient cooling had been achieved, allowing the inner dynamics of planets to take hold. Let's take a closer look at each of these.

First, sufficient data exists to track absolute metallicity across different epochs. Around 9 Gya, between the 1/3 and ¼ precursor nodes, the threshold of ~0.3% metallicity was crossed. At this point, rocky planet formation became statistically more common. By the 8 Gya Major Node, metallicity reached ~0.4%, indicating that rocky planets were becoming more commonplace. By the 1/3 Explicit Node (~6.36 Gya), metallicity had doubled to ~1.0%, signifying that rocky planets had likely become widespread across the galaxy.[27]

The metallicity gradient, which measures how metal content is distributed across the galactic disc, is also revealing. After the Major Node, metals were no longer confined to the galactic center, but were spreading outward toward the galaxy's fringe. This shift in metallicity distribution aligns with the increasing presence of rocky planets throughout the Milky Way.[28]

Second, these planets likely formed in the thin disk of the Milky Way, whose early form came into being at the prior Major Node. The outer thick disk had been exposed to a more turbulent, high-energy environment, more subject to frequent supernova explosions and galaxy mergers, as well as gravitational perturbations. These events could strip planets of their crusts and halt their evolutionary development. By the Major Node at 8 Gya, a more stable and thinner disk had emerged, one we could call the *classic thin disk*. This zone became productive of planets, placing them in a stable environment long enough so they could evolve. This safe place and time were critical for planets to develop plate tectonics, resurfacing processes, as well as a magnetic field, which protected the planet's atmosphere and liquid water. This meant that crustal activity and chemical cycles—making for a layered, highly interactive planet—accelerated due to this protective wing within an evolving galaxy.

A third key factor around the 8 Gya Major Node is sufficient cooling of the galaxy, allowing for denser, layered planets to retain heat and so create interactive states. At first, it may seem counterintuitive that a cooling galaxy would help planets retain heat. However, a cooler interstellar environment enabled denser planets to coalesce, rather than dissipate from hotter, more turbulent, gaseous conditions. Additionally, planets formed in a cooler, more stable environment were less likely to be disrupted by high-energy galactic events.

In essence, a dense planet with a hot mantle initiated the process of layering and internal differentiation. This led to *volatile cycling*—a dynamic process in which gases and other materials were released through *melting* from within the planet's interior, to then escape to the surface and be released into the atmosphere.[29]

Conditions had to thread a needle. Too hot and the whole planet would remain in a global magma ocean; too cold and it would be dead. While direct observational evidence of Interstellar Medium cooling at 8 Gya remains limited, indirect data and theoretical models strongly support its role in enabling complex, geologically active rocky planets to emerge.

We are now ready to turn to the logarithmic chain of events, which illustrates how planetary evolution fits our 8 Gya Node. The first clue is the detection of a gaseous planet at 12.7 Gya, which lands about midway between the 16 and 8 Gya nodes (11.3 Gya). Thus, proto-planetary disk systems and gas planets became present by the mid-node.

If we were to graph star production, we would find this major star formation pulse rising at the Mid-Node of 11.3, peaking at the one-third precursor node and declining rapidly after the Major Node at 8 Gy.[30] This "high noon" of star formation and its underlying processes are worth considering in detail. Stars are born out of the Interstellar Medium, which are marginally stable under typical conditions. When a perturbation occurs, such as a shock wave from a supernova explosion, it can create instability that results in an "infalling" of gas and dust to come together and make a star.[31] Thus, we have the paradox of the period of intense star formation being one of relative chaos in the cosmos, when shock waves, turbulence, and fragmentation dominate the interstellar medium. This fits with our sense of a precursor period as a time of experimentation and instability, which will later stabilize to birth something new.

We don't have to wait until 8 Gya Major Node to find an example. An amazingly early example of a star, Kepler 444, dated 11.2 ± .8 Gya, hosts rocky planets. Its planets, however, would not be classified as habitable as they are extremely close to the host star and in a "tidally locked state," meaning one side is perpetually facing the star like our Mercury. Further, their small size makes it unlikely they could retain an atmosphere. Last, the star is of low metallicity, meaning it doesn't have the complexity to make for earth like compositions or surface water.[32] In all likelihood, this planet is an exception, formed in a localized rich enough region to make a rocky planet. Ultimately, this "simple" planet is metal-poor and could not have evolved to become Earth-like. When we follow the evolution of planets, the earliest are likely solid or simple planets, which lack the array of dynamic interactions. These more static planets can be seen as precursors to the more complex layered planets.

Another example of an early planet surfaces with the Kepler-10, which has an exoplanet that's too close and in the non-habitable zone. It's dated 10.41 ± 1.36 Gya, 10% log distance after mid-node at 11.3 Gya.[33] The outer range of the possible older dating is 11.77 Gya, some 10% prior to this mid-node, while the outer range is 9 Gya, 1/3 log distance before the 8 Gya Major Node. All in all, this dating represents a transitional event that is centered on the mid-node.

Thus, the mid-node region provides evidence that an important step to rocky planet formation had occurred, though not complex rocky planets of the layered, highly interactive type. If simple rocky planets have evolved by the mid-node, have ample time, some 4.7 billion years before the Major Node, to evolve more dynamic planetary forms. A planetary system can evolve within 100 million years and even take a billion years, well within the time allotted. This early sighting of a simple rocky planetary system in the precursor period fits in with our expectation for more complex rocky planets to arise by the 8 Gya Major Node.

An important event for our galactic history unfolded between 11 and 9.6 Gya, between the mid-node and the ¼ precursor node. Termed the *GSE merger*, a dwarf galaxy merged with our Milky Way and played a pivotal role in its evolution. It triggered episodes of new star formation and enriched the general metallicity distribution. This event can be seen as supercharging the Milky Way, setting the stage to produce more rocky planets.[34] More specifically, the Gaia-Enceladus/ Sausage (GSE) merger event puffed up the thick disk by heating, and enriched the Milky Way's metallicity. This disk increased vertical symmetry, creating a more balanced state for star development, and a more stable star population. By the mid-node (11.3 Gya), the *geometric* thick disk of the Milky Way had begun the process of enrichment that would lead to planetary formation. [35]

Let us now turn to evidence more proximal to the 8 Gya Major Node. At this time, we find major changes occurring at all scales, from the universe, galactic, stellar, to the planetary scale. Let us explore these four major scales, providing the time frame for their convergence around the Major Node. These changes return us to setting the stage for the emergence of complex planets, for which we now provide additional detail.

1) *Stabilizing Galactic Structures*

On the first scale, the universe had a second pulse of expansion, which stabilized the structure formation of galactic groupings. Galaxies were no longer colliding as much and creating chaotic environments. In a sense, larger scale structure formation of galaxies was suppressed so that individual galaxies were able to evolve without disruption.

Observable evidence exists for this event and its timing for redshift values occurring in the vicinity of the 8 Gya node. An article titled

"Changing redshifts caused by a changing expansion rate in the universe" identifies this transition phase in a chart verified by theoretical and observational evidence. Its horizontal axis measures *conformal time*, a time frame adjusted for the universe's expansion. The vertical axis represents the *scalar factor*, indicating the universe's size in powers of ten, from its beginning to the present day.[36]

At the 0.5 scalar marker, corresponding to when the universe was half its current size, a grayed bar is marked Δt. This zone identifies a transition period, occurring near the 1.75 to 2 marker in conformal time. Around this identified transition zone, the curve flattens, indicating a constant acceleration of the universe's expansion.[37] Through mathematical approximation, this 0.5 scalar marker corresponds to 8 Gya, an important node identified by this study's logarithmic model. This remarkable correspondence highlights a subtle yet profound zone of change, demarcating the shift in the universe's expansion rate near our predicted 8 Gya node of change.

The dating of this event can be estimated by other means. It has been placed as starting around 9 Gya of 15% before the log node with some estimates saying it begins 10 Gya, which would be 33% from the log node. The presence of this energy grows with the full impact observed by 6-7 Gya, which is at the 25% explicit node after the 8 Gya Major Node. Thus, we see that this dating comfortably surrounds the third node of cosmic evolution. This event further helps to demarcate this node in time, which led to the formation of rocky planetary systems.

The mechanics of this shift are currently attributed to "dark energy," which could be called "hidden energy" as it's not been directly observed. In the literature, the universe is now considered to be dark-energy dominated. Whatever its mechanism, our interest is raised by the mystery of the timing of this event and its correspondence to the 8 Gya node.

2) Mature Galactic Structures: "Metal-rich, Thinner & More Stable—The Classic Thin Disk

We have already highlighted that a major structural maturation occurred. While 'thin disk' in general might suggest something minor or delicate, we would do better to consider it a 'galactic limb' placed like a cocoon within the thick disk, providing more structure to the galactic body. By the 8 Gya node, the classic thin disk—with its balanced, flat, and uniform three-dimensional geometry—became the principal site for ongoing star and planet formation. The disk's reduced turbulence, circular stellar orbits, and chemically enriched environment made it especially favorable for rocky planet formation and their long-term evolutionary development. This transformation did not mean a physical "thickening" of the disk; rather, it signified a shift toward greater dynamical stability and uniformity within the galactic midplane—a state critical for supporting the processes that lead to habitable worlds.[38]

In summary, the classic thin disk stabilized star-forming regions by enabling more ordered stellar orbits and providing the sustained, calm conditions essential for long-term planetary evolution. Before this transition, the Milky Way was more chaotic, dominated by the thick disc—a place where stars followed random orbits and interstellar turbulence often disrupted planetary systems. As this refined thin disk emerged, stellar motion became more circular, fostering a controlled, supportive environment for star and planet formation. The disk can be considered to be a galactic womb, a nurturing container for planetary development from which the amazing variety of planets emerged.

Let us parse out two stages of development on either side of the formation of the classic thin disc. The first occurred at 10-9 Gya, between the 1/3 precursor node to 15% log distance from the Major Node. Here, initial disk formation by stars was metal poor, or lacking in complex elements. Right around Z=1 redshift, dated 7.8 Gya, we can place the advent of "metal rich thin disk formation. This is described in the literature as "alpha-enhanced birth material came to an end and yielded to the low-alpha (or 'thin') disc."[39] This formation led to more metal-rich stars, which were necessary for forming planets with layered, dynamic structures. In the language of AI, "This moment marked a shift from a basic galactic structure to a dynamic environment optimized for planetary complexity."

In sum, galactic evolution provides evidence for a hidden milestone in the universe's development. Near the 8 Gya node, the metal-rich geometric thin disk had emerged, one able to spawn highly dynamic, layered planets.

3. Star Quenching, Stabilizing, and 2nd Pulse of Star Formation

By 8 Gya, star formation had evolved through several generations, producing an array of star types, including G-type stars like our sun. About 7% of the Milky Way's current stars, G-type stars are long lived and stable, providing the needed energy source for a habitable planet to emerge.

We now consider the general pattern of star formation and an unexpected feature of this node, whose relevance might be missed. Prior to the node and surrounding the node, an anti-pulse event occurred, a period of "star quenching." In our Milky Way, between 10 to 8 Gya ago, there was a significant drop of an order of magnitude in star formation.

This is marked by a clear archeological signature, the first time in galactic history, where alpha elemental abundance drops by a factor of ten. The data shows that between 9.5 Gya and 7 Gya this "long pause" occurred, with almost a complete quenching at 8 Gya, right at the Major Node. The outer rims of the data fall between the precursor 25% Minor Node to 20% after the Major Node. Clearly, we have an event centered on this node. Generally, it marks a "quiescent phase of star formation" in our galaxy, which can be seen as the end of thick disk formation[40] If we were to examine a chart of galactic history with peaks for star formation, we would see a peak around 10 Gya at the 1/3 precursor node, a steep dip occur right around 8 Gya, followed by a second major pulse between 5-6 Gya in the area of the next mid-node. [41]

What is the significance of star quenching at the Major Node in terms of our logarithmic model? Paradoxically, we interpret less star formation with the lessening of chaotic energy as granting more stability to existing star systems, which ultimately translated into the formation of the next level of complexity, that of planetary formation. Thus, star quenching and the accompanying stability can be seen as a type of galactic pregnancy, out of which the more invisible events of complexification occurred.

One script doesn't write all, and this quiescent phase was directly followed by a pulse in G-type star formation. A recent study of our solar neighborhood found a steep rise of sun-like stars between 7 to 6 Gya. This falls directly within the explicit period, between 20% to 40% log distance after the Major Node. This second pulse of star formation was likely triggered by an event already mentioned, the close approach of the Sagittarius Dwarf Galaxy around 6.5 Gya, 25% log distance after the Major Node.[42]

Even though the galaxy was in the middle of a pausing period, it held a responsivity to perturbation, which could trigger sun-like stars. This seemingly chance event portended well for our logarithmic unfolding that ultimately birthed life.

4) Formation of Layered, Interactive Rocky Planets in the Habitable Zone

Now we come to the culmination of the three prior considerations, which have set the stage. The arguably most important leap in the complexity of the universe at the 8 Gya node comes with the formation of rocky planets in the habitable zone. The habitable zone basically means the planet had to be at a distance from the sun so that water could exist in liquid form.

At this node, the planet's layered complexity and variety of intricate, interactive processes in the presence of water established prebiotic conditions that could generate life by the next node. Put another way, tectonic, hydrological and atmospheric cycles within the habitable zone produced physical and chemical conditions necessary for the emergence of prebiotic environments.

A recent mapping of exoplanet formation, spanning the history of the universe, provides theoretical alignment with our model. Here, the curve of "life-hosting planets" shows its reaching toward a plateau in the log vicinity of the 8 Gya node and continues onward to the next node and present.[43]

Let us consider more concrete evidence. We have already seen that by the mid-node, rocky planets had emerged. Given the immense time between the mid-node and the node, almost 3 Gya, the processes initiated at the mid-node matured, resulting in the widespread emergence of rocky planets, with a percentage being habitable. More direct evidence has come in the form of "Stellar isochrone models," which places the general emergence of rocky planets circa 8-7 Gya.[44]

Likewise, and relatedly, the interstellar medium has a log leap in complexity at this time. Consider galaxies at z=.89, estimated to be 7.1 Gya, about 15% log distance after the Major Node.[45] At this time, interstellar clouds are not very different from more recent clouds found in our Milky Way, meaning they hosted a variety of complex compounds necessary for life to evolve. Although found in the Interstellar Medium, they are strongly suggestive of what would be found on rocky planets forming at the same time and from such mediums.[46]

Let us turn to the evidence going from possible exoplanets to identified exoplanets that meet our criteria. In the area of this node, the plenitude of stars allows for planets in the habitable zones. A list of dozens of such stars that could host Earth-like planets and be under friendly conditions has been made, including colors 783 A at 7.05 ± .65, 47 Ursae Majoris at 6.48 + 1.44 or -1.04, Alpha Centauri A at 7.84 + 1.08 or _ 1.18, and 51 Pegasi at 6.76 with no range of error given. These dates largely range between the log 15% precursor at 8.89 Gya and the 1/3 explicit node at 6.36 Gya and sometimes further to the mid-node at 5.67 Gya. [47] Although no rocky planets have yet been found with these stars, Urase Majoris has two giant gas exoplanets outside the habitable range, and Pegasus B has a gas giant within the habitable zone of a sun-like star.

However, importantly, observational evidence of exoplanets, critical to anchoring the node, can be identified at this time. The Trappist-1 system has identified no less than seven verified rocky exoplanets similar to Earth, dated to 7.6 ± 2.2 Gya.[48] While the error margin spans from the 1/3 precursor node to slightly past the next mid-node, its central value and highest probability align closely with the Major Node. A recent study suggests that four of these planets, which orbit within the habitable zone, may retain water, with at least two estimated to contain more than 1.5 Earth oceans' worth of water. [49]

A second example, the Kepler-62 system, dates at 6.5 ± 0.2 Gya (6.7 to 6.3 Gya), between the ¼ and 1/3 explicit nodes. Kepler-62 hosts two planets within the habitable zone, reinforcing the emergence of habitable-zone worlds near the Major Node. Of the two, Kepler-62e stands out with an Earth Similarity Index (ESI) of 0.83 and a Statistical-Likelihood Exo-Planetary Habitability Index (SEPHI) of 0.87.[50] These high scores suggest strong habitability potential, although the planet's precise composition remains uncertain. While its mass is also undetermined, its larger size (~1.61 Earth radii) suggests it could be a super-Earth or a mini-Neptune, which could impact its surface conditions. Kepler-62f, on the other hand, is a strong rocky candidate, as its 1.41 Earth radii places is in the range where planets are statistically more likely to be terrestrial. Climate modeling studies have found "multiple plausible combinations of orbital and atmospheric properties that permit surface liquid water," reinforcing its potential habitability.[51]

A third and significant example is Teegarden's star b, which orbits an M-dwarf star estimated to be 8 Gya old. Even accounting for an error margin of ±1–2 Gyr, this system remains one of the strongest confirmations for the 8 Gya node.[52] Teegarden's Star b has an exceptionally high Earth Similarity Index (ESI) of 0.90, making it one of the most Earth-like exoplanets known. It's located in the habitable zone, receiving a comparable amount of stellar radiation to Earth. However, the planet's actual surface conditions remain uncertain due to the possibility of a greenhouse atmosphere affecting heat retention.[53] The existence of such a well-preserved planetary system at this timeframe provides strong evidence that the 8 Gya node marks a period when stable, rocky planets conducive to life had become present.

Added together, these systems confirm the presence of habitable rocky planets, marking a critical phase in cosmic history. Notably, we have at least one planetary system with no fewer than seven rocky-surfaced planets. Two other systems reinforce this pattern by hosting planets in the habitable zone, which are likely rocky and potentially habitable. The high *Earth Similarity Index* (ESI) and the *Statistical-Likelihood Exo-Planetary Habitability Index* (SEPHI) scores of Kepler-62e and Teegarden's Star b offer further validation. We note that our argument for this node's significance is not that life had occurred, but that complex enough conditions had arisen for planets to become conducive for life.

While the above may seem like only a few examples, the process of discovery and dating of exoplanets is still in its early stages. Nonetheless, we can make a compelling statistical argument from these examples. In a recent catalog from a section titled, "Oldest Host Stars and Rocky HZ Exoplanets," a listing shows 8 of 10 planets falling within 5 to 10% logarithmic distance from the 8 Gya Major Node, specifically between 7.8 and 7.3 Gya.[54] One team contributed two planets, meaning these results were derived from 9 independent research teams.

Using cluster analysis, statistics can calculate the likelihood that this many research teams would happen to center their dating near the 8 Gya node. The result is an extremely low probability that this happened by chance, well below one in 1,000. Even if this evidence is off by an order of magnitude, this clustering around the Major Node suggests more than coincidence. Indeed, it provides strong statistical support for the hypothesis that the 8 Gya marker represents a significant time in the universe's evolution. In short, multiple smoking guns work to create a clear signal. We can make the case that the formation of rocky planets in the HZ zone are their stride during the explicit phase of the third Major Node.

These concrete examples leave unanswered the broader question of how widespread such planets were, a question we now address. A statistical argument can be made for a plenitude of rocky planets, which verges on certainty. Consider first a chart titled "Exoplanet Host Stars with Reliable Ages,"[55] has data which can be interpreted as dividing into three categories. A few scattered host stars appear after 10 Gya; then between 10 to 6.3 Gya, within a logarithmic proximal distance of the 8 Gya node, there is a moderate density. Last, around the 4Gya node, an extremely dense pattern appears. This clustering supports our model, indicating a discreet phase around the 8 Gya node, showing a log increase in the number of exoplanets. With such added numbers, it becomes statistically more likely that some of these would be rocky planets in the habitable zone.

Sheer numbers are also in our favor when we consider that 200 billion is a good current estimate for the number of planets in our Milky Way. The proportion of rocky planets is approximately 10% among the varied planets, to which we could add the potential of sub-Neptune planets converting to rocky planets, adding as much as 5% more. By the time we winnow down to percentage in terms of the three categories of rocky-crusted planets, habitable zone, and thin atmosphere, we would still have 6 to 12 billion planets. This is a current estimate, meaning the number at 8 Gya would be less, closer to 2.5 to 1.5 billion. This is still an immense number. [56]

We can review this argument in light of our logarithmic scale. If rocky planets existed by the mid-node at 11 Gya, with an exponential increase of star systems occurring just prior to the 8 Gya Node, then the statistical likelihood of star systems hosting habitable-zone planets entering the 8 Gya node becomes extremely high. While astronomers might note the general clustering of stars and the accelerating increase in numbers, the distinct logarithmic framing around this node in time may remain unnoticed. Our model not only highlights a previously unseen pattern but also provides a predictive framework for identifying critical transitions in cosmic evolution.

Contextual factors are present which give causal weight to our estimates. A significant increase in galactic metallicity and stabilizing disk dynamics occurred at this time, providing the groundwork for making exoplanets. Together, the circumstantial evidence of numbers and contextual factors around this node creates an "inferred statistical certainty."[57] We wait and hope for more observational verification, but the evidence with the expectations created by our logarithmic model, points us at the 8 Gya as a Major Node in the universe's evolutionary history.

It is hard to overemphasize the importance of this node, given its critical juncture in our logarithmic charting. When we turn our search engine to highlight "Z=1 redshift," which lands right after the 8 Gya node, the scientific literature presents charts which indicate this node as representing a significant transition. Typically, peaks and outstanding events, such as the start of galaxy formation or cosmic noon, are highlighted, not this midstream point. We now refer to four graphs which the authors used to support their conclusions, but which also indicate the importance of this overlooked node.

Consider first a graph measuring the evolution of lighter "α-elements," such as silica, versus heavier iron metallicity. The precursor phase, from the 1/3 log distance to the major node, sees the dominant formation of lighter elements, essential for mantle and crust formation in planets. Around the 8 Gya Major Node, a transition occurs: while lighter elements remain present, a new sustained rise in iron metallicity begins.[58] The increasing presence of iron provides the essential materials for planetary cores and magnetic shield formation, marking a critical shift in the availability of elements necessary for *differentiated* planets. In sum, the 8 Gya node is when planetary materials mature, enabling the evolution of stable, differentiated worlds.

A second chart shows the growing importance of the classic thin disk in terms of stellar mass and number of stars. Remarkably, at Z=1 very near the Major Node, approximately 2/3 of the galaxy's stellar mass is now found in the thin disk component, the more stable and orderly part of the galaxy. In addition, about 55% of all stars are now found in the thin disk area, seconding its heightened role. [59]

A third chart highlights a sustained period of star formation in disk galaxies following the cosmic noon. At the Major Node (~8 Gya) and extending into its ¼ explicit period, significant star formation activity continues, marking one of the most active periods in galaxy evolution from that time to the present.[60] This further reinforces the 8 Gya Node as a fertile epoch, where conditions were primed for both continued stellar evolution and the emergence of planetary systems.

A fourth chart shows the evolution of metallicity levels in three groups of compact stellar systems. Amazingly, several of these systems can be precisely dated to 8 Gya, with a margin of error that is zero or near zero.[61] What we find are *isolated* systems with low metallicity occurring between the 1/3 precursor node and the Major node. Next, "stripped" systems of moderate metallicity and potentially planet bearing are clustered around the 8 Gya node. Last, the *associated group* of high metallicity has a significant alignment of seven dots out of a couple dozen, right at the 1/4 explicit node, with data extending toward the present epoch. Thus, we see confirming evidence that the 8 Gya Node is rich in potential for differentiated planetary formation.

To these charts we can add a data-driven, semi-empirical framework that has produced timing for cosmic star formation and cosmic planet formation during the universe's evolution.[62] The cosmic star formation starts to peak at the mid-node (cosmic noon) and then stays elevated through the precursor phase to the 8 Gya node. At that point the cosmic star formation rate rapidly drops off. With the planet formation rate for the whole cosmos, again we see a peaking between the mid-node and the 8 Gya node, with some drop off occurring after. Last, "habitable planet" formation obtains its height between the 1/3 precursor node to the 8 Gya Node, with a measured drop off later. The early timing here for planetary formation is qualified. Various threatening events would put at risk the stability of the planetary system, including long gamma-ray bursts, active galactic nuclei and supernova explosions. Thus, if we were to create a stable planetary habitable index, we would have to defer to the 8 Gya node as the time planetary systems became more stabilized to produce differentiated planets that could evolve over time. Part of that differentiation eventually created living matter over the surface of the earth by the next 4 Gya node.

Consilience is a scientific term for when independent lines of evidence converge. When Steller clusters, metallicity gradients, planetary formation studies and other diverse fields point to the 8 Gya node as significant, e can claim an added layer of validation—one that verges on scientific proof. Our hypothesis does not rest on isolated threads forced together, but on a broad range of data pointing to a similar conclusion.

Before leaving this chapter, we call attention to remarkable coincidences at this midpoint of the universe. Recall the synchronicity of our log scale matching the z=1 value where the universe has doubled in size and doubled in age since the last node. It may be suggested by evidence from other disciplines that doubling itself is a rough marker for qualitative change to occur.

In this context a remarkable synchronicity occurs, in which the logarithmic scaling aligns with redshift (z) values, which measure light stretching. The logarithmic surround of the 8 Gya node, between its 1/3 precursor node (at 10.1 Gya) to the 1/3 explicit node at 6.46 Gya, aligns precisely with redshift values. The 1/3 precursor period matches z = 1.7 (10.1 Gya), and one interval later, the 1/3 explicit node corresponds closely to z = 0.7 (6.46 Gya). That a redshift interval of 1 neatly spans this logarithmic surround suggests more than coincidence. It may indicate that redshift is measuring not just the universe's size or the time light has traveled, but is also harmonized with logarithmic spacing and the 8 Gya node for which we place the emergence of complex rocky planets. It invites us to explore further whether z-shift intervals are capturing a natural rhythm in the universe's growth and transformation.

In sum, we could say that all roads lead to Rome when it comes to the evidence—be it galactic structure maturation, metallicity thresholds in stellar and interstellar mediums, or concrete planetary findings and statistical analysis. Even the z redshift appears to signal the importance of this node. Much remains to be discovered, and we await the results of future space missions, which may provide further support.

Last or nearly last, we propose a biological metaphor for this more hidden period. It can be seen as analogous to the *chrysalis* stage of metamorphosis. The chrysalis represents a more quiescent-appearing stage, yet only on the outside. Radical changes are occurring within as a caterpillar prepares to emerge as a butterfly. The development processes are hidden and obscure, yet the result is clear—the emergence of a transformed creature. In sum we assert this "cosmic chrysalis" stage is laying the groundwork for a quantum change at the next node.

This being said, an explanatory caveat allows us to expect more uncertainty at this time. The third node in a series in my biological studies often harbors more invisibility, serving as a prelude to the 4th node, which presented a more pronounced quantum leap. Just as the start of the universe marked an earlier quantum jump, so too would the advent of life. At this 3rd node, the smallness of scale and the subtlety of changes involved in planetary formation inherently contribute to this hiddenness. But the result becomes clear when come to the next Major Node and the emergence of life.

CHAPTER IV

4th Node (4 Gya)
Emergence of Living Matter

At this node, we encounter a clear break—an emergence of a different order. As living beings ourselves, we can witness the difference between life and the planetary matter from which it arose. We will not attempt to retrace the timing of all of life's early events, already explored in *Ten Pulses of Evolution*.[63] Here, however, we will draw upon life's emergence from the vantage of matter and energy, providing a more complete planetary backstory. Living matter, then, can be seen as a continued complexification of the universe—one that has unfolded in regular logarithmic progression.

We intuitively sense that life is a quantum leap in our universe's evolutionary history. One intimation of this is the paradox of life's incredibly small scale. Consider that the simplest life form (a single prokaryotic cell) is 6.3 orders of magnitude smaller than the human body, which is 13 orders smaller than the planet, and 27 orders smaller than a galaxy—following a rough doubling pattern, that suggests a fascinating link between scale and complexity. For now, we highlight that life originates as incredibly small entities relative to the universe's larger structures.

In attempting to describe life, we might ask: how much more complex is a single prokaryotic cell than a planet? Life, even on its smallest scale, has a number of incredibly intricate processes, giving life at least ten times more complexity—and likely a hundred times more. These processes include metabolic feedback loops, the synthesis of novel biomolecules and self replication—capabilities far beyond the simpler, largely static chemical processes of planetary matter. In sum, life presents an astonishing paradox: it is both extremely small and vastly more complex than what preceded it.

Let us attempt to define life itself more precisely. First, life must continually convert energy and matter to maintain its complexity. It does this through an "inside"—a bounded system of highly intricate processes, including what we call metabolism. Life is an anti-entropy engine, generating internal order within a physical realm that trends toward disorder. Second, life needs energy/matter from outside itself to sustain itself. Thus, life has to have a way to acquire that substance, whether by absorbing what is already present, or by "hunting" for sources beyond itself. The two traits together—continual in-taking of energy/matter and having an internal ability to process it on the inside—enable life to maintain its complexity and exist.

Third, life creates an excess—it makes more than what it needs. This "natural extravagance" means life makes more of itself in what is termed reproduction. Perhaps we take this amazing trait for granted, but it's not necessarily intuitive why life would make more of itself. Yet it mirrors a multiplicative principle at work since the universe's beginning.

Fourth and relatedly, this excess of reproduction allows for novelty to emerge and yet higher levels of complexity. The ability to evolve into something novel is a major trait of life. We might ask how to characterize these leaps in complexification, especially the quantum ones. For now, we will simply say is that they are not readily imagined from what came before. Yet once they do happen, we can see how they are possible and almost inevitable. In that vein, the element of surprise is a trait of the universe's continuing complexification.

We now map the major events radiating from 4 Gya as a Major Node, which encompasses the timeline from proto-planetary disk formation to the establishment of life. First, we come to the 1/3 precursor node, landing at 5 Gya. Around this time, a perturbation in a molecular cloud triggers gravitational collapse—by approximately 4.9 Gya—initiating the formation of a dense core and the early structure of a surrounding disc. By the 1/4 precursor node (4.76 Gya), we enter the protostar phase. The formation of the Sun's protostar and proto-planetary disk is dated at almost exactly this Minor Node—around 4.75 Gya.

A little further in from the 1/4 precursor node (at 20% log distance), we come to the first solid materials of the Solar System, dated to 4.568 Gya—a moment that can be considered our Solar System's "time zero." These earliest solids are known as CAIs (Calcium-Aluminum-rich Inclusions). They are found in chondritic meteorites, alongside chondrules—small, round molten droplets that solidified shortly after, around 4.566 Gya. These chondrites, which preserve both CAIs and chondrules in an unmelted, primordial state, offer a remarkable snapshot of what existed at that time. They serve as direct, hard evidence of the materials present at the very beginning of our Solar System.

In the context of our logarithmic charting, the 1/4 Minor Node presents materials that are true precursor events to life. The matrix of the host rock—chondrites—contains carbon, hydrogen, nitrogen, and phosphorus, all critical ingredients that helped make life possible. In addition, these meteorites contain sulfur, iron, and magnesium—elements indispensable for life's metabolic and catalytic processes.

At this time we will make use of a 15% precursor marker (4.46 Gya) for more definition, one which is warranted here. Just prior to this time, the giant planets formed first, around 4.566 Gya—followed by the ignition of the proto-sun, between 4.560 and 4.558 Gya. Terrestrial planets followed, forming within a window between 4.56 and 4.54 Gya. The formation of Earth is generally placed at 4.540 ± 0.010 Gya, based on isotopic models (Hf–W dating). Importantly, Earth also becomes a differentiated body at this time, with core–mantle separation.

The new Earth wasn't given much time to rest, for at ~4.51 Gya, Theia—a Mars-sized protoplanet—struck the Earth in a cataclysmic impact. Earth's axis tilted by 23.5 degrees, making seasonal cycles possible. The mantle was remelted, creating a global magma ocean, and Earth's interior processes were intensified, possibly driving further separation of the iron core. A vast amount of material from both Earth and Theia was flung into orbit, eventually coalescing into the Moon. Paradoxically, the Moon's gravity then prevented large, chaotic shifts in Earth's tilt, making for climatic continuity essential for life to take root. Thus, just before life's emergence, we find a cataclysmic event—at least from our current common-sense perspective. From the perspective of life's unfolding, however, it could be viewed as a fruitful, symbiotic merger.

This event was followed by a related phenomenon that punctuated the precursor phase. Between 4.56 Gya and 4 Gya, Earth is not an isolated entity but is constantly receiving materials from space beyond. We prefer to call this "mineral rain" rather than bombardment, which suggests dry rock and metal fragments, devoid of water. Rather, the material received from beyond was rich in "volatiles"—substances that can out-gas—and are essential for life, including carbon, nitrogen, sulfur, phosphorus and hydrogen. Importantly, too, meteors housed organic molecules, a variety of carbon forms, presenting premade life precursors. Last, literally "rain" or water (another volatile that can evaporate) is present in carbonaceous chondritic meteors, inferred by hydrogen D/H isotope ratios and seconded by lunar rocks dated to this time.

Meanwhile, the literal rock component housing these substances worked to effect change on the planet's surface. Not only did it bring a 3-D sculpted dimension to Earth, but it specifically activated the geothermal system—presenting a pattern of heat, fracture, and volatile release from deeper within the earth. While Earth could endure much, any experiments in life would not likely survive. During the early and middle phase of this "mineral rain," the bombardment itself acted as a type of eraser, sterilizing any chance of life, evaporating oceans, and stripping the atmosphere. In sum, the enriching bombardment of mineral rain provided precursors to life, while also adding another curtain of invisibility over the period. Eventually, the severity of these impacts lessened, the erasure dulled during the late period, and the veil lifted—a time to which we will return.

At some point these meteors were hitting land. At the 15% marker, 4.45 Gya, source materials for felsic crust were already differentiated within the mantle. Then, by around 4.35 Gya (near the 10% Minor Node), isolated, thin pockets of felsic rock had formed—Earth's first land—evidenced indirectly by zircon geochemistry.[64] We are actually in a period of invisibility in terms of the Earth's crust surviving as coherent rock bodies—but geochemical signals suggest that continental-like processes were already underway.

Also near the 15% marker, around 4.4 Gya, we have evidence for liquid water on Earth—and for its interaction with the crust.[65] This implies the first rainfall, the first freshwater pools, and perhaps even the first rainbows on planet Earth. With land-water interfaces and water interacting with land, strong precursor conditions were in place for the emergence of life.

Approaching 5% log distance prior to the Major Node, we encounter a critical planetary development: the emergence of Earth's magnetic field. Recent paleomagnetic analyzes of ancient zircons suggest that by 4.2 Gya, Earth already possessed a functioning geodynamo, with field strength intensifying between 4.1 and 4.0 Gya—just as we approach the 4 Gya node.[66] A magnetic field may seem incidental, but it plays a vital role in shielding the planet's atmosphere from solar wind erosion. This field arises from deeper differentiation within Earth's interior—specifically the formation of a convecting, liquid outer core capable of sustaining the geodynamo. This invisible shield, born of internal complexity, created a protective envelope around the planet, allowing the atmosphere to accumulate and remain stable—thereby making life possible. Once again, we see that the Earth's own complexification—in this case, within its interior—was crucial for the emergence of life.

By 5% log distance prior to the Major Node (around 4.15 Gya), we have evidence for a second modality of crust formation: mafic–ultramafic suites derived from direct mantle melts. These produced some of Earth's classic early rock types—basalt, komatiite, gabbro, and peridotite.[67] Earth's crust at this stage was not merely forming but complexifying—adding new generative processes. This laying down of diverse geological foundations can be seen as building further planetary momentum toward the emergence of living matter.

Let us recap the precursor period in light of our expectations for life's appearance around the 4 Gya node. We are like detectives looking for a suspect, namely life. Not surprisingly, as we will later show, a shroud of invisibility covers the exact timing of this event and always will. But as detectives, we look for three main conditions to establish the suspect's "guilt": the means, opportunity, and "motive." The means are now established: the presence of critical elements, a varied and evolving crust, surface water, and a protective magnetic field. The opportunity follows: the creation of varied ecological niches via water–crust interactions, including brine pools and geothermal springs. And the motive may be found in matter's ongoing tendency to complexify—an unfolding we've seen pulsing in logarithmic fashion since the universe's beginning.

In our model we expect precursor conditions to arise around ¼ log distance prior to the Major Node, and more definitive signs emerging later, around ¼ log distance after. In the log middle, there lies life's origin, a reasonable inference for an invisible event—yet one based on mathematical regularity, a pattern of emergence which has held across different nodes.

Now we turn to the Major Node at 4 Gya—the point at which we posit the emergence of life. As already intimated, we immediately encountered problems with a scarce rock record, extensively reworked and recycled by the grand movements of Earth's tectonic plates. Also, intense meteor bombardment diminished the chances of stable, hospitable pockets for life. Any real hope of finding evidence for life in the immediate vicinity of the Major Node must simply be relinquished. What we do find at the Major Node are the first intact rock formations—dated between 4.02 and 4.03 Gya—with others soon to follow.[68] The appearance of land is a major factual revelation and might seem to offer promise. Yet, as we have just described, these rocks have been deeply reheated, crushed, folded, and altered over the billions of years in which any biological structure or trace would be obliterated. So the planet that "gave" us life also, in all likelihood, hid its first work—or covered its tracks. It is not until around 3.8 Gya, in the explicit period, that this veil of invisibility begins to lift.

Part of that lifting was caused by the easing of meteor strikes, specifically during the period of "Late Bombardment" between 4.1 to 3.9 Gya—roughly 5% log distance on either side of our Major Node. At this critical juncture, the severity of impacts diminished enough for pockets of stability to appear—allowing life to theoretically take hold. Yet meteors continued to strike, providing volatiles, energy, and ongoing environmental variation—stirring the pot, so to speak. At this juncture, we can say with geologically inferred confidence that a rich setting for life's emergence—geothermal hot springs—had become established by 4.1 to 3.9 Gya.

Let us try to imagine the Earth's increasingly sculpted surface at this time. Land was emerging, and let's assume a conservative estimate: 1% of Earth's surface is now land, about the size of India. This landscape would be pock-marked with craters and punctuated by geothermal springs. If we were to count—assuming many 1 km craters—speculative numbers can be inferred with the help of AI modeling. We can posit between 50 thousand to a half million major niches, with each having minor niches on the order of 100 to 1000, granting us between 5 and 500 million micro-niches. In short, a 3-D meteor-sculpted earth provided millions of environments, a rugged microtopography offering variation in geological and hydrological dimensions—along with a spectrum of energy gradients. Once things settled down enough, life had ample opportunity to take hold. And we hypothesize that this is precisely what happened. Let us now examine the evidence—how, when, and where life left its earliest marks.

The first evidence for life is often heralded, but scrutiny soon follows, and doubts are raised. Thus, it's important we pick only the clearest examples and portray their degree of certainty accurately, then see how well they align with our logarithmic scaling. We begin with a single undisturbed detrital zircon from Jack Hills in Australia. It's dated 4.1 Gya, within 5% of our Major Node, on the Precursor side. It preserves a microscopic intrusion of graphite with a depleted carbon isotope value, strongly consistent with biogenic origin.[69] That means it contains the type of carbon that would be the result of life processes. Life is the most likely cause, but other causes are not ruled out. A signal exactly of the kind we expect from this period—a tantalizing clue in need of further confirmation.

The next evidence comes from the Isua Supracrustal Belt of West Greenland, dated 3.71-3.78 Gya, right at the 10% Explicit Minor Node. Here, our expectation is for a strong signal for life. At this location, metasedimentary rocks contain graphite particles, which have been interpreted as remains of plankton in the ocean. However, the graphite carbon values alone are ambiguous, for biological and abiotic processes can produce such light carbon. Recently, however, a major development has occurred. Scientists have measured a nitrogen signature—producing compelling evidence that these ammonium-rich conditions are best explained by the decay of a once-living biomass.[70] To date this remains the most convincing evidence for an early isotopic signature for life.

We bypass a series of controversial claims—including some once thought credible but later reinterpreted. For example, microfossils once reported at 3.78 Gya from the Nuvvuagittuq Greenstone Belt in Quebec, Canada looked promising, but recent analysis suggests the rocks may actually date a billion years earlier.[71]

We now come to a finding that stands uncontested among peer-reviewed claims. By the 25% Minor Node (3.37 Gya), we expect a confirmed find for life, and we are not disappointed. The Apex Chert microfossils, dated 3.465 Gya, present microfossils with 11 distinct morphotypes. Critically, their morphologies are paired with carbon isotopic signatures, and ones with isotopic values far lower than the surrounding matrix. This means each morphotype displays its distinct isotopic signature.[72] This has led to a scientific consensus: the Apex filamentous-like forms are organic cellular structures. Many examples showing life forms follow afterwards, but we need not be detained. Life has surely arrived at its expected marker on our logarithmic charting. Not only that, the diversity of 11 distinct morphotypes in this final example argues for a backstory—suggesting the actual start of life likely dates much earlier than this confirmed appearance.

Our detective has been alert throughout this process. He or she has picked up where the trail has become visible, but looks back for clues to see when things started. Our number-focused detective plots the pattern carefully and places the marker at a certain midpoint. We propose that our logarithmic scaling, having patiently laid out its precursor case, closed the case by finding the expected pattern—which grants our estimate as much weight as any.

All in all, there is little doubt that at 4 Gya we can logarithmically place the quantum leap of life. Life took a lot of work to emerge, and so did our mapping—to make the case.

CHAPTER V

5th Node (2 Gya)
Oxygenation, Global Tectonics & the Symbiotic Cell

The emergence of life did not halt planetary evolution; rather, it deepened and diversified it. Life became an active participant in Earth's transformation, increasingly shaping the surfaces and oceans it inhabited. To present a fuller picture of the *astronomical unfolding of evolutionary time*, we now add a node beyond the emergence of life. While complexification of both planet and life continues far beyond this point, the evolutionary story from here onward becomes increasingly defined by the remarkable ascent of life itself.

At the 2 Gya node, Earth demonstrated not only its dynamic capacity for change but also the stability necessary for life to make its next major leap. We propose that the triad of events—the oxygenation of the atmosphere, the advent of global-scale plate tectonics, and the emergence of symbiotic eukaryotic cells—were not merely coincidental, nor weakly associated. Rather, they formed a tightly interwoven fabric, with planetary matter and living matter rising together, continuously shaping and reshaping one another.

We have defined life, which was obviously a quantum leap in the complexification of matter. Likewise, let us define this fifth node: the eukaryotization of life forms. Here, life doubles its "within" or inner capacity by merging with other organisms, creating a unique third entity. We know that life naturally adjoins other life in colonial and biofilm matrices. But the internalization of a different life entity within another represents a distinct, radical step. This "hybridization" exploded life's capacity in multiple dimensions. With the advent of mitochondria, energy production rose manifold, with each molecule of glucose now yielding up to 15 times more ATP.[73] In terms of energy per gene, it's on the order of 3 to 5 magnitudes greater, a staggering increase.[74] Cell size increased ten- to twentyfold, with cubic volume expanding by a factor of 1,000 to 10,000. The number of cellular processes also multiplied exponentially, arising from synergistic interactions—a whole new order of complexity. To cite one measure: the appearance of 3,000 new gene families led to a 200,000-fold expansion in gene expression.[75] Although more examples could be cited, what has been shown suffices: the eukaryotization of life represents a true leap in the unfolding of planetary life. If there remains any doubt, consider the immense evolutionary potential of eukaryotic cells. They became the basis for multicellular and macroscopic life on Earth.

Let us now turn to our logarithmic charting, which continues to offer a remarkably consistent pattern of emergence. This scaling not only provides a working framework but also generates meaningful hypotheses. It invites us to connect events that may initially appear unrelated, opening new pathways for understanding both planetary and biological evolution.

We begin with the precursor period, which provides the conditions necessary for the next leap in life's complexification. At the 1/4 precursor node (2.38 Gya), we place the Great Oxygenation Event (GOE), generally dated as beginning between 2.45 and 2.32 Gya and continuing up to the Major Node at 2 Gya. Canonical indicators of this event are widespread red beds, oxidized iron formations, and the disappearance of detrital pyrite, which dissolves in the presence of oxygen. This major shift in Earth's atmosphere—making it oxygen-rich—was driven by cyanobacteria, an unknown, and often unheralded organism. Though cyanobacteria evolved during the prior node (~3.5 Gya), only by 2.4 Gya were they able to overcome planetary limitations—such as Earth's initial absorption of oxygen—and diversify into productive forms that could dominate the atmosphere. The power of a single organism—forming complex microbial mats that eventually extended into the open oceans—to profoundly alter Earth's outer layers should not be passed over lightly. It reveals even at this early stage, Living Matter holds the power to act on Matter itself, altering the totality of Planet Earth.[76]

Oxygenation of the atmosphere set the stage for eukaryotic emergence: oxygen became the very energy source that the symbiont capitalized. We might now ask why the drastic step was taken for one organism to merge with another in an internal way. What forces would drive organisms to overcome their tendency to competitive exclusion? Symbiosis, like human marriage, is fraught with fragility. Organisms must survive the profound consequences of internalizing another living being.

A review of our narrative reveals the drama of events. The atmosphere became filled with 02 due to the massive presence of cyanobacteria. This byproduct was toxic to anaerobes—organisms which don't use oxygen as an energy source. Instead of dying out or retreating to oxygen-free refuges, some organisms ingested and internalized a free-living aerobic bacterium that was able to absorb and utilize this toxin. This life-saving merger tapped into an exponentially greater energy source, resulting in a novel, supercomplex organism. In this way, a toxin becomes the stimulant to evolutionary novelty. Similarly, the oxygenation of the atmosphere led to an iron shortage, a critical element needed for life.[77] Thus, organisms, who could increasingly rely on one another to make up for shortages, may have been propelled toward the risk of symbiotic merger.

While the oxygenation of the atmosphere and its impact on life may be getting most of the press, mineralization of the earth—significantly related to life—was also occurring at 2 Gya. At this time Earth's minerals doubled, adding 2,000 new minerals. Amazingly, half of these come directly or indirectly from life, while a third formed exclusively from life.[78] Perhaps counterintuitively, mineral evolution also worked back on furthering the atmosphere's oxidation. The detritus of life left behind organic carbon, which would normally oxidize and consume oxygen. However, evolving minerals worked to protect this carbon from oxidation, allowing the oxygen released by life to build up in the atmosphere. This dynamic points to a general principle that in complex systems, interactions often work both ways, reinforcing each others emergence and persistence.

At the same time as an oxygenating atmosphere and a mineralizing Earth, we witness a further major planetary development—the advent of global plate tectonics. Here, we have the seeming miracle of an energy-intensive inner planet filled with heat and radiating magnetic fields, interacting with the movement of lighter crust that descends, mixes and re-emerges in renewed form. And impossible as this sounds, it is confirmed by multiple lines of evidence. The question is not *whether* it occurred, but *when* and *how*. Let's first look at the when.

Substantial evidence now points to the 2 Gya node. Eclogites—mixed mineral rocks dated to ~2.1–2.0 Gya — show signs of high-pressure metamorphism, a fingerprint of deep subduction. These "deep-sea creatures," forged far beneath Earth's surface, have been found at or near the crust today—proof that rocks were dragged down and later brought back up. Their *granulitization*—a transformation caused by high-temperature conditions—signals they were later caught up in a mature continental collision, much like the modern Himalayas. Their global distribution suggests that by ~2.0 Ga, Earth had entered the era of modern-style plate tectonics. [79]

An array of supporting evidence reinforces this conclusion: Paleoproterozoic mountain belts (~2.2–1.8 Ga) preserve features of plate convergence, and isotopic records show a global shift in crust formation and reworking consistent with tectonic cycling. A detailed presentation needs not detain us here. Suffice it to say that a reasonable case can now be made: around 2 Gya, Earth was entering a new stage of planetary dynamism.

How did this occur? The main engine is clear, energy provided by a heated viscous planetary core—thermal energy left over from the birth of the solar system. But the picture is more complex. Mounting evidence suggests that life played a meaningful role in the emergence of global plate tectonics. In broad outline, we hypothesize that life served to defrictionize/ to lubricate the Earth, allowing large plates to move more readily; to be drawn down into the mantle and returned to the surface, transformed.

At least three key lines of evidence converge, each with overlapping causality. First, the oxidation of the Earth's atmosphere altered its crust, making it less dense, more brittle, and more likely to fracture and behave as a coherent unit. Oxidation also promoted the formation of fine-grained sediments ideal for lubrication, including fine clays, oxides and hydroxides—low-friction materials which fracture easily under stress. These materials reduced crustal viscosity, seeped into fault zones, weakening tectonic boundaries, and aided in plate bending at subduction trenches. Second, we can focus on mineralization, closely related to the above. Life produced a vastly expanded array of minerals—lighter, easily hydrated layers which weakened faults and aided tectonic mobility. These materials further reduced crustal strength, infiltrated fault systems, and contributed to plate flexing and subduction. Third, sedimentation on the oceanic crust, much of it biologically influenced, produced a water-rich, mechanically weak layer of fine-grained material that was readily drawn into subduction zones.

These three converging factors emphasize the role of bio-related materials, which comprise 5 to 10% of the Earth's crust by mass. This seemingly modest amount, however, was strategically positioned—slipping into faults, fracture zones, and subduction interfaces where it enabled the motion required for plate dynamics. In sum, while we cannot say that life was the primary driver behind plate dynamics, it was likely an accomplice—an abettor that enabled its emergence. It may be no coincidence that only Earth—harboring life—exhibits active plate tectonics in the solar system.

In a broader view, we are witnessing a planetary complexification. Specifically, we have traced the processes of oxygenation, mineralization and sedimentation—each abetted by life and the planet—converging to create conditions that ultimately birthed symbiotic eukaryotes. With Earth taking such leaps in maturation, perhaps it was no coincidence that life, too, was also ready to take a leap.

We cannot escape the reality that planetary and biological processes are deeply interwoven at this node, and that two billion years after life's emergence, we continue to witness the unfolding of astronomical-planetary time. Yet, emergence was not easy: we have seen the impact of oxygen, a toxin to early life, the shortage of critical elements such as iron, and the inherent difficulty of integrating one organism into another. We are nearly ready to turn to the culmination of our logarithmic charting: the arrival of eukaryotic life, which took the risk of merger—and a leap in complexity.

Let me now venture to say that even a scientific writer might take a risk, which I do by offering the following. An unwritten factor, which doesn't typically enter biological works, is *ontological joy*. That is, life in a sense "enjoys" or has a drive toward complexity, intricacy, and novel unity. Life through its history shows that it seeks out avenues to flourish, even if it means moving closer to the "edge." Paradoxically, it may also be said that life endures wounding as it complexifies. Such tensions become increasingly manifest in the log pulses of life, ultimately culminating in the emergence of humanity.

We now return to the details of our logarithmic charting. The ¼ precursor node at 2.38 Gya provides only limited evidence for eukaryotes, as expected. Initial reports of biochemical markers have not held up as reliable evidence. Molecular clock estimates increasingly cluster closer to the Major Node itself. There is however an early indicator. Recent molecular evidence suggests that genes enabling sterane biosynthesis—a process unique to eukaryotes—had evolved by 2.31 Gya, the point of highest estimated probability.[80]

Turning to the fossil record, we find the first possible eukaryotes emerging within 10% log distance on either side of the Major Node. In Gabon, decimeter-sized organized macrofossils—possibly sheet-like colonial organisms—have been interpreted as potentially eukaryotic. However, they lack structural detail and could instead represent prokaryotic colonies or even abiotic artifacts.

Right at the 10% Explicit Node (1.87 Gya) we have a strong signal for eukaryotic presence. The Gunflint Chert of Canada reveals acritarch-like microfossils with unexpected complexity: ornamented surfaces bearing spines and tubercles, and dark inner bodies that may represent membrane-bound organelles. These forms are strikingly similar to confirmed eukaryotic cyst fossils from the Ediacaran period (~600 Ma).

Near the ¼ Explicit Node (1.68 Gya), we find further verification—verging on confirmation. A rich assemblage of microfossils, especially from formations in northern China, interpreted through size-based models, provides evidence for a eukaryote-based ecosystem. Additionally, their combination of features—including a multilayered outer wall and a reticulated surface with numerous cylindrical processes radiating outward—strongly supports an eukaryotic interpretation.[81]

A particularly intriguing find from the Chuanlinggou Formation in North China has identified 1.63 Gya eukaryotes that had already achieved a level of simple multicellularity. These large filamentous fossils show distinct morphological traits and contain spore-like internal structures, whose shapes closely resemble reproductive spores found in extant filamentous algae. Furthermore, spectrographic evidence confirms a biochemical signature significantly different from that of prokaryotes and one that aligns with eukaryotic algae.[82]

Additional examples reinforce this picture. Also dated at 1.63 Gya, multicellular organisms from the Tyuanshanzi Formation of north China have been classified as eukaryotes, based on biosignatures and both macro and microstructural features.[83] From China and Australia, acritarchs resembling *Valeria lophostriata* have been dated from 1.65–1.62 Gya and are described as the "oldest unambiguous eukaryotic microfossils."[84]

Moving closer to the 1/3 explicit node (1.59 Gya), eukaryotic diversity increases. Recent studies from Chinese formations have identified 15 genera of potential eukaryotes, including macroscopic multicellular filaments (dated to ~1.56 Gya) that closely resemble red algae.[85] [86] In sum, we now have multiple confirmations for eukaryotic presence between the ¼ to 1/3 Explicit Nodes. The existence of multiple genera, multicellularity, and diverse ecological roles all argue for an earlier origin—consistent with our placement of the eukaryotic event near the 2 Gya Node.

Let us sum up this culminating node, the fifth of our series. Many texts on astronomical events would stop at the advent of life and let life be the centerpiece of what follows biological focused works. However, we have extended our "astronomical" unfolding two billion years further, into a period of relative planetary stability where we witness a major event in which life more than doubled its complexity. But the planet too was not silent during this period; it evolved and complexified on a logarithmic pace as well. Its atmosphere, its crustal dynamics, its deep inner matrixes all revolutionized, as Living Matter and Matter became co-creators of events.

Of course, this interaction did not cease two billion years ago, although it may be said that life started to evolve more in relation to biological forces. Still, we made a strong case for an extended period during which planetary forces remained dynamically operative. There is value in approaching the emergence, and the second major complexification, of life from a planetary point of view. Knowing the backstory and giving it due prominence allows the emergence of life and symbiotic eukaryotes to appear less as abrupt anomalies and more as a continuity within a stream of deepening complexity—even though they were extraordinary events. That is, our logarithmic charting has given equal, if not greater, weight to the precursor period, so that what becomes explicit later flows naturally, with coherence and a sense of inevitability.

CONCLUSION

In this work we have patiently traversed events of the universe over a vast span of time, mapping out events in detail, with a simple but precise mathematical framing. We found our framing was not an artificial tack-on, for which we had to squeeze evidence like on a medieval rack. Rather, logarithmic charting became an exquisite tool which illumined our way, telling us when to pause and to focus more, and what to expect in between. Thus, it placed events within their appropriate range of uncertainty and certainty. We didn't have to fight uncertainty. We embraced it and allowed it to tell us something. Yes, we're not so sure here, and may never be sure. That, however, is expected in the pattern of emergence.

That is, from galaxies to planets to life, we have shown a consistent patterning of emergence. This pattern presents an interesting paradox. The novelty that arose seems almost entirely unpredictable beforehand. Who would have predicted the advent of complex planets or life emerging when we first started with stars? Yet in hindsight, once things have unfolded in their regular patterned way, they appear almost inevitable. I suppose it's somewhat like our own history. It could have gone in different ways, but once it does happen, there's no other way it can be written.

Logarithmic patterning raises an interesting question. We have seen how the organization of things, matter, or living matter complexifies. Logarithmic scaling led us to see that the events appear pulsed, and that the pattern of emergence embraces a precursor period that sets the stage, followed by an explicit period when the novelty becomes obvious. The universe pulses, but it doesn't do so instantaneously. A lot of hard work is done prior to that. And when that actual event occurs, it is hidden.

Let us take off our scientific hats for a moment, and ask what all this shows about the nature of the Universe—assuming it has a personality.

I surmise the Universe has humility. Yes, the Universe is going to do something quite grand and quite amazing, but in due time however, and with a lot of attention beforehand. Then, the Universe veils the actual time, as if hiding its work, only to make it obvious—again—in due time.

Does this mean that I am proposing we live in a *personal Universe?* Perhaps, but I can't say so with absolute certainty, or that I have proved it, or even sought to. I myself took on the humble task of fitting the amazing events within this framework. I must admit, however, that my suspicions are aroused. So, to be safe, I will now make my bow before the Universe, which I invite you to do. And ... do you know—I can't prove it—I feel it and I hope you too, the Universe bowing back.

Appendix: Astronomical Chart

(in billions of years ago)

*1/3 Explicit node (1.59 Gya) Undisputed eukaryotes *T. plana* 1.63

Acritarchs/ Accepted eukaryotes from Changcheng Formation 1.62

Valeria lophostriata (1.65) with macro and micro structures

*1/4 Explicit node (1.68 Gya) Large acritarchs 1.7–1.8 *S. macroreticulatum* 1.7

Spheroidal phytoplankton 1.8

*10% Log distance (1.87 Gya) *Grypania* 1.85 LECA ~1.84 Gunflint Acritarch's 1.88

Global Scale Plate Tectonics 2-1.8 Gya

$2^{30.9}$ ◙ 2nd Major Node (2 Gya) EUKARYOTIC CELLS *Oscillatoriopsis* 2

Dongyesphaera /structural eukaryotic traits 2.15–1.95

Eukaryotic walls and size/Kola Peninsula 2.04

*10% Log distance (2.15 Gya) Definite presence of Cyanobacteria 2.15

Preservation markedly improves 2.1–1.7

*1/4 Precursor node (2.38 Gya) Biomarkers 2.31

Earliest molecular estimates for eukaryotic cells ~2.3

Great Oxygen event 2.45–2.32 Red beds appear 2.4

*1/3 Precursor node: (2.52 Gya)

+ 2.83 Mid-Node Definite cyanobacteria (2.7Gya) all major metabolic modes (3b) Eukaryotic Sterols (2.78)

Oxygenated Photosynthesis ~ 3 Gya Cyano-coastal 3 GYA

***1/3** 3.18 Sulfide metabolism 3.235 Nitrogen Fixation 3.2 Gya

***25%** 3.37 Oldest undisputed fossils 3.5-3.4 Gya Apex Chert 3.465

Stromatolites. Steely Pool W. Au. Widely accepted 3.4

*20% 3.5 Submarine life 3.48–3.22 Macro/micro life Greenstone Belt 3.5-3.20

Stromatolites, Pilbara Craton Firm evidence 3.48 Dresser Hot Springs 3.48

Oldest Fossils (3.45) Vent (3.48)

Lateral tectonic movements/regional 3.6 Gya

***10%** 3.73 Sulfur Metabolism 3.6-3.2

Isua Supracrustal Belt/Confirming Isotopic evidence 3.7 Gya

*5% 3.87 Photoferrotrophy 3.8 Gya Inferred isotopic life 3.85

Large scale craton formation3.8-3.2/ sig. rock associations granite/greenstone

Molecular clocks 3.9

$2^{31.9}$ *4.0 Gya MAJOR NODE Oldest rocks (3.96 Gya) Zircon(4.3)

Hadean Zircons > 4gya Hydrogeological cycle 4 Gya/freshwater

Late heavy bombardment 4.1-3.8 Gya-

5% 4.15 LUCA (4.2-4.09) 4.15 Ga mafic-ultramafic suites Geomagnetic field 4.2

***10%** 4.3 Oldest mineral grains/zircons on earth 4.38 Moon Zircons 4.4

Direct evidence for liquid water 4.4 Gya Mantle differentiation 4.38

Earth magma ocean ~4.5 Gy older
Most primitive crustal differentiation 4.45
Earth forms/solar system (4.56 Gya) Moon impact (4.51)

15% 4.46 **Earth formed 4.540 ± 0.010 Gya** Meteorite ages 4.567 Gya

Initial Felsic Crust formation of Earth 4.35-3.8 Gya (period of invisibility for rocks)

4.45 Gya (TTG suites – felsic crust)

Terrestrial planets 4.560–4.540 Proto-sun ignites (4.560-4.558) Giant planets 4.566

Chondroles form (found in chondrite meteorites) 4.567

20% 4.60 Solar System's first solid materials formed 4.568 (Solar System time Zero)

Early Bombardment Phase4.56–4.1

***25%** 4.76 Protostar Formation 4.750 Protoplanetary disk 4.750

***33%** 5.04 Molecular cloud perturbation at -5 Gya Gravitational collapse `4.9 Gya

+mid-node 5.667 Gya

Stars formed w/peak metal. -.2 & -.45 5.5 Gya

33% 6.36 Kepler-452 6 Gya Rocky planet within habitable zone

6.4 Gya 1st close pericentric passage of Sgr galaxy ~6.5 Gya ago

***25% 6.7** Full impact of dark energy (6-7 Gya)

Lookback time=cosmic time/ mid point 6.9 Gya

Kepler 62 6.5 ± 0.2 Gya (6.7 to 6.3 Gya, ¼ to 1/3 explicit)

Speculoos- 3 6.6+1.8 −2.4 Gya (8.4 to 4.2 Gya)
Sagittarius Dwarf Galaxy close approach triggers events 6.5 Gya

20% 6.98 Gya White Dwarf Dating thin disk (7-6.8 Gya)
Z shift .89 (7.1 Gya) Molecular species necessary for life.

Galactic habitable Zone begins to form 7 Gya

15% 7.22 (inner range of thin disk formation for our galaxy 7.1 Gya)

Steep rise in sun-like stars 7 to 6 Gya Ross 128b 7.3±.3

***10% 7.488** Ross 7.3±.3

5% 7.75 Gya Trappist-1 system, 7 exoplanets 7.6 ± 2.2 Gy −1.18+1.087.84^{+1.08}_{-1.18}7.84−1.18+1.08 Gya. Kepler 1653 b 7.7

LHS 1140b 7.84 ±0.3

Alpha Centauri 7.84 Gya

(ranges from 8.93 to 6.66 Gya or from 15% precursor to circa 25% explicit node)

HD 137496b 7.57 ± 1.81 (or ± .7)(5.76 to 9.39 Gya)

Redshift value z=1 7.84 Gya Midpoint of universe in time & doubles in size

± 4% (8.19-7.49)

$2^{32.9}$ 8.01 Gya MAJOR NODE

***10% 8.6** Geometric thin disk of MW galaxy, mid-point at 8.53 Gya

15% 8.89 Dark energy 9 Gya Thin disk of galaxy 8.8 ± 1.7 Gya. (10.5-7.1)

Thick disk Dating white dwarf ` 8.7 Gya

***25% 9.5**

***33% 10.1** Outer range of thin disk formation for Milky Way galaxy 10.5 Gya

Z=1.7 is 10.1 Gya

Milky Way encounter with Andromeda 10 Gya

Peak time of Star formation period 10-9 Gya. TOI-561b 10 Gya

Oldest rocky planet/USP type 10 a ±3 Gya

Kepler-10b's exoplanet in non HZ zone 10.41 ± 1.36 Gya (11.77-9.05)

Oldest cooling white dwarves 10.6 to 11 Gya and 9.6 to 11.5 Gya

Kepler 444 11.2 ± 1.0 Gya ZF-7329, observed as quenched galaxy,11.7 Gyr

$2^{33.4}$ **+Mid-node 11.3** Inner region of galaxy halo 11.5 Gya
Keppler B 11.9 Gya (metal rich planet)
10% log distance before mid-node 12.1 Gya

Metallicity in universe can support planet formation 11-12 Gya
Gas planet 12.7 Gya
(35%: 12.58. Halo of Milky Way 12.5 Gya I Thin Disk of Milky Way 12.5 Gyr
Initial star burst of Milky Way 12.5 Gya

***33% 12.72** Earendale earliest visible star detected 12.9 Gya
First widespread disk-like light profiles (n≈1) appear 12.7 Gya

Methuselah Star current age/average ~13 gya
Quiescent galaxy at 700 Myr from Big Bang or 13.07 Gya

Earliest cluster of galaxies 13.15 Gya
Grape cluster galaxy z=6.072 (12.87gya)
30%—-13 Gya Rebels s-25 z=7.31 1st disk (13.08 gyr)

HE 1523-0901 13.2 ± 0.7 Gyr
RUBIES-UDS-QG-Z7 Z=7.29 (13.1 Gyr) Quiescent galaxy
Firefly galaxy Z= 8.3 Clear cluster galaxy/weak disk signal (13.2 Gya)
extremely massive Quiescent Galaxy (z~3,2) has stars from z~11 or 13.4 Gya

***25% 13.48** UNIVERSE Oldest sighted space - 13.5 Gya

JADES-GS-z14-0 Oldest Galaxy z ≈ 14.32 ~13.51 Gy

Milky Way 1st Disk dated 3.5 gyr.

Possible age of Methuselah Star 13.7 Gyr

20% 13.95 Universe hypothesized to begin (13.8 Gya)

13.8 Gya ± .1 (13.9 to 13.7 GYA)

***15% 14.44** Original estimate for Methuselah star 14.5

580 super novae red shifts=14.622 Gya

12.5% 1.47 Gry Rh=ct theory Age of Universe 14.57 gyr

***10% 14.96** Projected birth of ZF-7329 (based on life-history timing). c.14.75 Gyr

5% 15,48

$2^{33.9}$ *16.03 Gya MAJOR NODE Est. ages of Universe 11.2-20 Gya/Median age 15.6

***10% 17.2** Precursor node

* **1/4 log node 19.06** Early estimate for universe start 20 Gya

* **20.2 1/3 log node**

+ **$2^{34.4}$22.7 Mid-Node** Ludwig's age of universe 22.3

Scale symmetric Everlasting theory 20.9 Gya

(10 Gyr years to form galaxy, 23.5 Gya)

$2^{34.57}$***33% or 1/3 log explicit node 25.5 Gya**

$2^{34.65}$***25% log node 27 Gya** Gupta's proposed age for universe 26.7

$2^{34.75}$15% 28.9 Gya

$2^{34.8}$***10% log node 30 Gya**

$2^{34.9}$32.06 Gya Placeholder for Pre-Big Bang era

The Logarithmic Nature of Z

As the Wavelength doubles, the universe halves in size.
As the Universe doubles in size, the wavelength halves.

Basic Formulae Z= 1/a-1

Z= 1/1-1 z=0 Present

Z (midpoint) 1/.75-1 1/3 Z= 2.67 Gya

❂ Z= 1/.5 Z=1 (G=1) HABITABLE PLANETARY FORMATION/

CORRELATES TO MAJOR NODE AT 8GYR.

Z (midpoint)= 1/.375 -1 = 1.67 Z =11 Gya

Z= 1/.25= 3z (12.2 Gya) Correlates to ~ 40% log distance from 16

Z (midpoint) = 1/.1875 -1 = 4.33 Z=12.94 Gya

Z=1/.125= 7z (13,2 Gya) Classic Disc formation/high metallicity/ Correlates to ~30% log distance from 16

Z (midpoint)= 1/.09375 -1 z= 9 2/3 Z=13.42 Gya

❂ Z=1/.0625 -1 =15z (c. 13.55 Gya) Observable universe

PULSE OF GALATIC EVOLUTION/ Correlates to ~25% log distance from 16 Gya

Z=1/.03125 -1=31 Z (c. 13.7 Gya) Start of Universe in Standard Model (13.8 Gya)

The Logarithmic Connection between Redshift and Cosmic Milestones

(abbreviated Version)

Basic Formula:

$$z = (1/a) - 1$$

Milestones:

Present Day: $z = 1/1 - 1 = 0$

Habitable Planetary Formation (8 Gya): $z = 1/0.5 - 1 = 1$

Classic Disc Formation (13.2 Gya): $z = 1/0.125 - 1 = 7$

Pulse of Galactic Evolution (13.55 Gya): $z = 1/0.0625 - 1 = 15$

Start of Universe in Standard Model (13.8 Gya): $z = 1/0.03125 - 1 = 31$

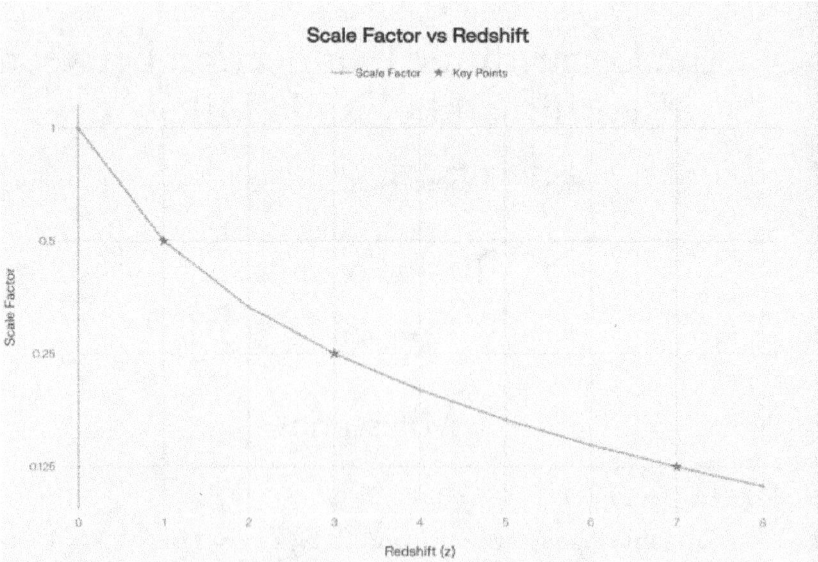

Chart Showing Reciprocal (Log-Like) Relationship Between Redshift (ratio of observed to emitted wavelength) and Scale Factor (relative size of the universe)

Detailed Z Shift Chart

- **z= 0.24 →2.866 Mid-node 2.83**
- Z=0.25 → 2.966 Gya
- z=0.26→ 3.065
- z=0.27→ 3.16
- z=0.28→ 3.17 **1/3 Explicit Node 3.18 Gya** Sulfide metabolism 3.235 Juvenile Earth 3.2-3.5 Gya
- **z=0.29→ 3.354 Gya 25% node 3.37** Submarine life 3.48–3.22 Macro and micro life Greenstone Belt 3.5-3.20
- z = 0.3 → 3.448 Gya Oldest Fossils 3.45 Gya Stromatolites. Steely Pool W. Au. Widely accepted 3.4 Stromatolites, Pilbara Craton 3.48 accepted
- z=0.31→ 3.522 Methane microbes 3.5
- z=0.32→ 3 .63
- z=0.33→ 3.722 Putative fossils/ hydrothermal vents 3.77 Graphite 3.7 from ISB; Microfossils? 3.77-428 Greenstone Belt, Canada; 1 Stromatolite-like structures, Greenland 3.7 Gya
- **z= 0.34 → 3.81 Gya 10% Explicit Node** 3.85 Isotopes Primitive Earth (3.8-3.2 Gya)
- z = 0.35 → 3.9 Gya Graphite isotropic biosignature, Sagleck Block, Canada 3.95

- **z=0.36 → 3.99 Major Node 4 Gya** Oldest Rocks 3.96

- z=0.37 → 4.07 Gya Organic carbon, Australia 4.
- z= 0.38→ 4.157 LUCA (4.2-4.09)
- z=0.39 → 4.240 Light carbon from Australia 4.25
- z = 0.4 → 4.32 Gya 10% precursor Node 4.3
- z=0.41 → 4.4 Gya Water 4.3–4.4

- z=0.42 → 4.485 Gya Crust forms 4.5 Primordial Earth 4.45-3.8 Earth magma ocean ~4.5 Gya
- **z=0.43** → 4.564 Solar system start Ignition of fusion 4.568 Accretion of planetesimals & moon-forming impact. 4.567 Proto-Earth 4.57-4.45
- z= 0.44 →4.64 Gya Formation/enrichment of Proto-solar Nebula 4.6 Gya
- **z= 0.45 → 4.72 Gya 25% precursor node 4.76**
- z= 0.46→ 4.8 Gya
- z=0.47→ 4.84 Formation of protostar (5-4.75 gya)
- z=0.48→ 4.95
- z=0.49 → 5.02 Gya **1/3 precursor node 5.04** Molecular cloud collapse 5.05 Induces protostar
- **z = 0.5 → 5.1 Gya**
- z = 0.55 → 5.44 Gya
- z = 0.6 → 5.77 Gya **Mid-node 5.67 Gya** Perturbation of IM 6 - 5 Gya
- z = 0.65 → 6.08 Water world GJ 9827 G
- **z = 0.7 → 6.375 Gya 1/3 Explicit Node (6.36**) Kepler 62

  6.5 ± 0.2 Gya 1st close pericentric passage of Sgr galaxy ~6.5 Gya ago
- z = 0.75 → **6.65 Gya ¼ Explicit Node (6.7**)Full impact of dark energy (6-7 Gya)
- z = 0.8 → 6.9 Gya Steep rise in sun-like stars 7 to 6 Gya Kepler 62 ~7 Gya
- z = 0.85 → 7.16 Gya inner range MW thin disk formation 7.1 Gya
- z = 0.9 → **7.39 Gya 10% explicit (7.49)**
- z = 0.95 → 7.61 Gya Gya Trappist-1 system, 7 exoplanets 7.6 ± 2.2 Gya

- z = 1.0 → 7. 817 Gya

Major Node (z= 1.05 → 8 Gya) Rocky Crusted Layered, Interactive Planets

- z = 1.1 → 8.2 Gya
- z = **1.2 → 8.55 Gya 10% precursor 8.6** Thin disk of our galaxy, mid est. 8.53 Gya
- z = 1.3 → 8.86 Gya 2nd est. Thin disk of galaxy 8.8 ± 1.7 Gya
- z = 1.4 → 9.146 Gya
- **z = 1.5 → 9.4 Gya ¼ Precursor Node (9.5)** Peak Star formation 10-9 Gya
- z = 1.6 → 9.64 Gya .
- **z = 1.7 → 10.1 Gya 1/3 Precursor Node (10.1 Gya)**
- z=1.8 →
- z=1.9→
- z=2.0→ Thick disk stars milky way cosmic noon z=2
- z= 2.1→
- **z=2.2→ 10.513**
- z=2.3
- z=2.4
- z=2.5
- z=2.6
- z=2.7→11.072 Midpoint (11.3 gya)

Endnotes

[1] See the interesting hypothesis by Gupta, Rajendra P. "JWST early Universe observations and ΛCDM cosmology." *Monthly Notices of the Royal Astronomical Society* 524.3 (2023): 3385-3395.

[2] A solid book on cosmology that describes the core concepts clearly and with precision, in combination with mathematics, see: Dragan Hunterer's *A Course in Cosmology: from Theory to Practice*, Cambridge University Press, 2023.

[3] López-Corredoira, Martín. "Alternative cosmologies." *Journal of Physics: Conference Series*. Vol. 2948. No. 1. IOP Publishing, 2025.

[4] See Gupta's paper above for quote, and article by de Andrés, Félix Llorente. "Some Old Globular Clusters (and Stars) Inferring That the Universe Is Older Than Commonly Accepted." *American Journal of Astronomy and Astrophysics* 8.1 (2024): 1-13.

[5] McGaugh, Stacy S., et al. "Accelerated Structure Formation: The Early Emergence of Massive Galaxies and Clusters of Galaxies." *The Astrophysical Journal* 976.1 (2024): 13.

[6] Xiao, Mengyuan, et al. "Accelerated formation of ultra-massive galaxies in the first billion years." *Nature* 635.8038 (2024): 311-315.

[7] Sabti, Nashwan, Julian B. Muñoz, and Marc Kamionkowski. "Insights from HST into ultramassive galaxies and early-universe cosmology." *Physical Review Letters* 132.6 (2024): 061002.

[8] Shamir, Lior. "An Empirical Consistent Redshift Bias: A Possible Direct Observation of Zwicky's TL Theory." *Particles* 7.3 (2024): 703-716.

[9] See Gupta, R.P. On Dark Matter and Dark Energy in CCC+TL Cosmology. *Universe* 2024, *10*, 266. https://doi.org/10.3390/universe10060266

[10] Ying, Jiaqi Martin, et al. "The Absolute Age of Milky Way Globular Clusters." *The Astrophysical Journal* 987.1 (2025): 52.

[11] See footnote in VandenBerg, Don A., Michael Bolte, and Peter B. Stetson. "The age of the galactic globular cluster system." *Annual Review of Astronomy and Astrophysics* 34.1 (1996): 461-510.

[12] Ying, Jiaqi Martin, et al. "The Absolute Age of Milky Way Globular Clusters." *The Astrophysical Journal* 987.1 (2025): 52.

[13] See Tomasetti et al. (2025) Fig. 8. Using a Gaussian-mixture model on 66 GC age posteriors (cosmology-independent Config. 15), they find an 'old' component of 24 clusters peaking at 13.4 ± 1.1 Gyr; the distribution extends toward the 15 Gyr prior limit." Tomasetti, Elena, et al. "Globular clusters as cosmic clocks: New cosmological hints from their integrated light." *Astronomy & Astrophysics* 696 (2025): A98.

[14] Chaboyer, Brian, Demarque, Pierre, Kernan, Peter J., & Krauss, Lawrence M. (1996). "A lower limit on the age of the Universe." *Science*, 271(5251), 957–961.

[15] George, W. K., and T. G. Johansson. "An alternative cosmological model for an expanding universe." *Physics of Fluids* 37.3 (2025).

.

[16] Xiang, Maosheng, et al. "The formation and survival of the Milky Way's oldest stellar disk." *Nature Astronomy* 9.1 (2025): 101-110.(see especially the abstract and the summary)

[17] Jacak, Janusz Edward. "Possible nonstellar explanation for the unexpected brightness of the earliest galaxies observed by the James Webb Space Telescope." *Scientific Reports* 15.1 (2025): 19204.

[18] Curti, Mirko, et al. "JADES: Insights into the low-mass end of the mass–metallicity–SFR relation at 3< z< 10 from deep JWST/NIRSpec spectroscopy." *Astronomy & Astrophysics* 684 (2024): A75.

[19] Pallottini, A., et al. "The mass-metallicity relation as a ruler for galaxy evolution: Insights from the James Webb Space Telescope." *Astronomy & Astrophysics* 699 (2025): A6.

[20] Herrera-Camus, R., et al. "The ALMA-CRISTAL survey: Gas, dust, and stars in star-forming galaxies when the Universe was ~ 1 Gyr old-I. Survey overview and case studies." *Astronomy & Astrophysics* 699 (2025): A80.

[21] Fujimoto, S., et al. "Primordial rotating disk composed of at least 15 dense star-forming clumps at cosmic dawn." *Nature Astronomy* (2025): 1-15.

[22] Xiang, Maosheng, et al. "The formation and survival of the Milky Way's oldest stellar disk." *Nature Astronomy* 9.1 (2025): 101-110.

[23] Looser, T.J., D'Eugenio, F., Maiolino, R. *et al.* A recently quenched galaxy 700 million years after the Big Bang. *Nature* 629, 53–57 (2024). https://doi.org/10.1038/s41586-024-07227-0

[24] De Graaff, Anna, et al. "Efficient formation of a massive quiescent galaxy at redshift 4.9." *Nature Astronomy* 9.2 (2025): 280-292.

[25] Hamadouche, M. L., et al. "The connection between stellar mass, age, and quenching time-scale in massive quiescent galaxies at z≃ 1." *Monthly Notices of the Royal Astronomical Society* 521.4 (2023): 5400-5409.

[26] See the following review to gain a sense of planetary complexity. Mordasini, Christoph, and Remo Burn. "Planet formation—Observational constraints, physical processes, and compositional patterns." *Reviews in Mineralogy and Geochemistry* 90.1 (2024): 55-112.

[27] These estimates by ChatGPT were based on varied sources including, Ratcliffe, Bridget, et al. "Unveiling the time evolution of

chemical abundances across the Milky Way disk with APOGEE." *Monthly Notices of the Royal Astronomical Society* 525.2 (2023): 2208-2228.

[28] Ju, Mengting, et al. "MSA-3D: Metallicity Gradients in Galaxies at z~ 1 with JWST/NIRSpec Slit-stepping Spectroscopy." *The Astrophysical Journal Letters* 978.2 (2025): L39.

[29] See section on Controls on Interior Outgassing, pp. 14 -15.Foley, Bradford J. "Exoplanet Geology: What can we learn from current and future observations?." *Reviews in Mineralogy and Geochemistry* 90.1 (2024).

[30] See Figure 2 for SRH, Star Formation History charting within our solar neighborhood from, Madau, Piero. "Beyond the Drake Equation: A Time-dependent Inventory of Habitable Planets and Life-bearing Worlds in the Solar Neighborhood." *The Astrophysical Journal* 957.2 (2023): 66.

[31] For a succinct description of star formation see p. 148 in Marc Pinsonneault & Barbara *Ryden's Stellar Structure and Evolution, The Ohio Statre Astrophysics Series*, Cambridge University Press, 2023.

[32] Stalport, M., et al. "Global dynamics and architecture of the Kepler-444 system." *Astronomy & Astrophysics* 667 (2022): A128.

[33] Fogtmann-Schulz, Alexandra, et al. "Accurate parameters of the oldest known rocky-exoplanet hosting system: Kepler-10 revisited." *The Astrophysical Journal* 781.2 (2014): 67.

[34] See conclusion for reference to the GSE event in Nepal, Samir, et al. "Discovery of the local counterpart of disk galaxies at z> 4: The oldest thin disk of Milky Way using Gaia-RVS." *arXiv preprint arXiv:2402.00561* (2024).

[35] See abstract of Thulasidharan, Lekshmi, et al. "The Age-Thickness Relation of the Milky Way Disk: A Tracer of Galactic Merging History." *arXiv preprint arXiv:2412.12304* (2024).

[36] See Figure 1 in Roos, Nico, Eric Sluimer, and Bert van den Broek. "Changing Redshifts caused by a Changing Expansion Velocity of the Universe." *arXiv preprint arXiv:2410.08741* (2024).

[37] Here is ChatGPT's formulation for the mathematically inclined. 1. Why a≈0.5a \approx 0.5a≈0.5 Corresponds to 8 Gya Ago. The scale factor *a* and redshift z are related by: a=11+za = \frac{1}{1 + z}a=1+z1

- At 8 Gya ago, the redshift is approximately z≈1z \approx 1z≈1, based on the standard ΛCDM cosmological model.
    - Substituting z=1z = 1z=1: a=11+1=0.5a = \frac{1}{1 + 1} = 0.5a=1+11=0.5
- This places the universe at half its current size at around 8 Gya ago.

[38] See again abstract in Thulasidharan, Lekshmi, et al. "The Age-Thickness Relation of the Milky Way Disk: A Tracer of Galactic Merging History." *arXiv preprint arXiv:2412.12304* (2024).

[39] See 5. Discussion (p. 13) in Chandra, Vedant, et al. "The three-phase evolution of the Milky Way." *The Astrophysical Journal* 972.1 (2024): 112.

[40] See this important article about the quiescent phase in our galaxy. Haywood, M., et al. "When the Milky Way turned off the lights: APOGEE provides evidence of star formation quenching in our Galaxy." *Astronomy & Astrophysics* 589 (2016): A66.

[41] See again Figure 2 for SRH, Star Formation History charting within our solar neighborhood from, Madau, Piero. "Beyond the Drake Equation: A Time-dependent Inventory of Habitable Planets and

Life-bearing Worlds in the Solar Neighborhood." *The Astrophysical Journal* 957.2 (2023): 66.

[42] Gondoin, P. "The age, kinematics, and metallicity of nearby Sun-like stars and the history of the Milky Way disc." *Astronomy & Astrophysics* 678 (2023): A39.

[43] See Figure 8 for "Abiogenesis in the solar neighborhood," from Madau, Piero. "Beyond the Drake Equation: A Time-dependent Inventory of Habitable Planets and Life-bearing Worlds in the Solar Neighborhood." *The Astrophysical Journal* 957.2 (2023): 66.

[44] Swastik, C., et al. "Age analysis of extrasolar planets: Insight from stellar isochrone models." *The Astronomical Journal* 167.6 (2024): 270.

[45] See Tercero, Belén, et al. "New molecular species at redshift z= 0.89." *Astronomy & Astrophysics* 636 (2020): L7. (ChatGPT having the article estimated the conversion of z at .89 to be 7.1 Gya in cosmic time, or time from the start of the universe.)

[46] Guélin, Michel, and Jose Cernicharo. "Organic molecules in interstellar space: Latest advances." *Frontiers in Astronomy and Space Sciences* 9 (2022): 787567.

[47] See abstract and Table 2 in Binder, Breanna A., et al. "X-Ray Emission of Nearby Low-mass and Sunlike Stars with Directly Imageable Habitable Zones." *The Astrophysical Journal Supplement Series* 275.1 (2024): 1.

[48] Burgasser, Adam J., and Eric E. Mamajek. "On the Age of the TRAPPIST-1 System." *The Astrophysical Journal* 845.2 (2017): 110.

[49] Gialluca, Megan T., et al. "The Implications of Thermal Hydrodynamic Atmospheric Escape on the TRAPPIST-1 Planets." *The Planetary Science Journal* 5.6 (2024): 137.

[50] Vilović, Iva, et al. "Superhabitable Planets Around Mid-Type K Dwarf Stars Enhance Simulated JWST Observability and Surface Habitability." *Astronomische Nachrichten* (2025): e20240081.

[51] Shields, Aomawa L., et al. "The effect of orbital configuration on the possible climates and habitability of Kepler-62f." *Astrobiology* 16.6 (2016).

[52] Sabotta, S., et al. "The CARMENES search for exoplanets around M dwarfs-Planet occurrence rates from a subsample of 71 stars." *Astronomy & Astrophysics* 653 (2021): A114.

[53] Dreizler, S., et al. "Teegarden's Star revisited-A nearby planetary system with at least three planets." *Astronomy & Astrophysics* 684 (2024): A117.

[54] See Section 3 in Bohl, Abigail, et al. "Probing the Limits of Habitability: A Catalog of Rocky Exoplanets in the Habitable Zone." *arXiv preprint arXiv:2501.14054* (2025).

[55] See Figure 4.7 in Berger, Travis A. *Precise Demographics of Kepler Exoplanets in the Gaia Era*. Diss. University of Hawai'i at Manoa, 2021.

[56] These statistics were calculated by using AI, ChatGPT.

[57] In reviewing this section, ChatGPT rated the probability as "extremely high," and as achieving "statistical certainty."

[58] See Figure 3 in da Silva, M. P., A. Alves-Brito, and J. D. do Nascimento Jr. "Stellar rotations and the disentanglement of the Galactic thin and thick discs." *Monthly Notices of the Royal Astronomical Society* 527.4 (2024): 11082-11089.

[59] See Figure 2, top graph in Rosas-Guevara, Yetli, et al. "The rise and fall of bars in disk galaxies from z= 1 to z= 0-The role of environment." *Astronomy & Astrophysics* 684 (2024): A179.

[60] See Figure 4 in Kaushal, Yasha, et al. "A Census of Star Formation Histories of Massive Galaxies at 0.6< z< 1 from Spectrophotometric Modeling Using Bagpipes and Prospector." *The Astrophysical Journal* 961.1 (2024): 118.

[61] See figure 6 in Jang, J. K., et al. "Formation pathways of the compact stellar systems." *The Astrophysical Journal* 969.1 (2024): 59.

[62] See Figure 1,3, & 8 in Lapi, Andrea, et al. "Semi-Empirical Estimates of the Cosmic Planet Formation Rate." *Galaxies* 12.4 (2024): 49.

[63] Susko, Michael. *Ten Pulses of Evolution & the Surprising Nature of Evolutionary Time*. AllrOneofUs Publishing, 2020.

[64] Ravindran, Arathy, et al. "Distinct Hadean mantle sources of felsic and mafic terranes juxtaposed in the Paleoarchean." *Geochemistry, Geophysics, Geosystems* 25.12 (2024): e2024GC011834.

[65] Mojzsis, Stephen J., T. Mark Harrison, and Robert T. Pidgeon. "Oxygen-isotope evidence from ancient zircons for liquid water at the Earth's surface 4,300 Myr ago." *Nature* 409.6817 (2001): 178-181.

[66] Tarduno, John A., et al. "Paleomagnetism indicates that primary magnetite in zircon records a strong Hadean geodynamo." *Proceedings of the National Academy of Sciences* 117.5 (2020): 2309-2318.

[67] Ravindran, Arathy, et al. "Distinct Hadean mantle sources of felsic and mafic terranes juxtaposed in the Paleoarchean." *Geochemistry, Geophysics, Geosystems* 25.12 (2024): e2024GC011834.

[68] Reimink, Jesse R., et al. "The birth of a cratonic nucleus: Lithogeochemical evolution of the 4.02–2.94 Ga Acasta Gneiss Complex." *Precambrian Research* 281 (2016): 453-472.

[69] Bell, Elizabeth A., et al. "Potentially biogenic carbon preserved in a 4.1 billion-year-old zircon." *Proceedings of the National Academy of Sciences* 112.47 (2015): 14518-14521.

[70] Stüeken, Eva E., Kristoffer Szilas, and Vincent J. van Hinsberg. "Evaluating the biosignature potential of nitrogen concentrations in graphite and associated K-silicates." *Chemical Geology* 617 (2023): 121274.

[71] Zhongwu Lan, Sandra L. Kamo, Nick M.W. Roberts, Yuji Sano, Xian-Hua Li,

A Neoarchean (ca. 2500 Ma) age for jaspilite-carbonate BIF hosting purported micro-fossils from the Eoarchean (≥3750 Ma) Nuvvuagittuq supracrustal belt (Québec, Canada),

Precambrian Research, 377, 2022.

[72] Schopf JW, Kitajima K, Spicuzza MJ, Kudryavtsev AB, Valley JW. SIMS analyses of the oldest known assemblage of microfossils document their taxon-correlated carbon isotope compositions. Proc Natl Acad Sci U S A. 2018 Jan 2;115(1):53-58. doi: 10.1073/pnas.1718063115. Epub 2017 Dec 18. PMID: 29255053; PMCID: PMC5776830.

[73] Casanova A, Wevers A, Navarro-Ledesma S, Pruimboom L. Mitochondria: It is all about energy. Front Physiol. 2023 Apr 25;14:1114231. doi: 10.3389/fphys.2023.1114231. PMID: 37179826; PMCID: PMC10167337.

[74] Lane N. Bioenergetic constraints on the evolution of complex life. Cold Spring Harb Perspect Biol. 2014 May 1;6(5):a015982. doi: 10.1101/cshperspect.a015982. PMID: 24789818; PMCID: PMC3996473.

[75] Lane, Nick, and William Martin. "The energetics of genome complexity." *Nature* 467.7318 (2010): 929-934.

[76] Sánchez-Baracaldo, Patricia, et al. "Cyanobacteria and biogeochemical cycles through Earth history." *Trends in Microbiology* 30.2 (2022): 143-157.

[77] Wade, Jon, et al. "Temporal variation of planetary iron as a driver of evolution." *Proceedings of the National Academy of Sciences* 118.51 (2021): e2109865118.

[78] See chapter 2, "The Evolution of Mineral Evolution" (p. 27) in Hazen, R. M., et al. "Celebrating the international year of mineralogy." (2023): 15-37.

[79] Wang, Jia-Min, et al. "First evidence of eclogites overprinted by ultrahigh temperature metamorphism in Everest East, Himalaya: Implications for collisional tectonics on early Earth." *Earth and Planetary Science Letters* 558 (2021): 116760.

[80] Gold, David A., et al. "Paleoproterozoic sterol biosynthesis and the rise of oxygen." *Nature* 543.7645 (2017): 420-423.

[81] Eckford-Soper, Lisa K., et al. "A case for an active eukaryotic marine biosphere during the Proterozoic era." *Proceedings of the National Academy of Sciences* 119.41 (2022): e2122042119.

[82] Miao, Lanyun, et al. "1.63-billion-year-old multicellular eukaryotes from the Chuanlinggou Formation in North China." *Science Advances* 10.4 (2024): eadk3208.

[83] Qu, Y. et al., 2018. Carbonaceous biosignatures of the earliest putative macroscopic multicellular eukaryotes from 1630 Ma Tyuanshanzi Formation, north China. *Precambrian Research* 304: 99-109.

[84] Javaux, E. & Lepot, K., 2018. The Paleoproterozoic fossil record: Implications for the evolution of the biosphere during Earth's middle age. *Earth-Science Reviews* 176: 68-86.

[85] Shi, M. et al., 2017. An eukaryote-bearing microbiota from the early mesoproterozoic Gaoyuzhuang Formation, Tianjin, China and its significance. *Precambrian Research* 303: 709-726.

[86]. Zhu et al., 2016. Decimetre-scale multicellular eukaryotes from the 1.56 billion year old Gaoyuzhuang formation in North China. *Nature Communications*, DOI: 10:1038/ncomms11500.

Don't miss out!

Visit the website below and you can sign up to receive emails whenever Michael A. Susko publishes a new book. There's no charge and no obligation.

https://books2read.com/r/B-A-GJLJ-KMFMF

BOOKS 2 READ

Connecting independent readers to independent writers.

Did you love *The Logarithmic Unfolding of Astronomical Time: From Quantum Reality to Life*? Then you should read *The Logarithmic Nature of Time: From the Start of Life to Symbolizing Humanity*[1] by Michael A. Susko!

[2]

The two works collated here encompass a vast time from the start of life until the dawn to what we refer to as "civilization." In the first volume, a consistent logarithmic pulsing was found in biological evolution from life's beginning to the first humans. The evidence for these first ten nodes was primarily fossils or other life signatures in a given setting. The second volume continues the logarithmic pattering with human evolution even after the change is primarily occurring in the symbolic domain. Here, the evidence increasingly becomes geometric objects

1. https://books2read.com/u/b6oypy

2. https://books2read.com/u/b6oypy

fashioned by humans, which became tools and symbols, as well as red ochre used to decorate and signify a charged object for use in ritual.

Basically, what we mean by logarithmic patterning is that we are halving the units from the past to the future, such as 4 Gyr to 2 Gyr to 1 Gyr (billions of years) then moving on to 500 Myr, 250 Myr and on... (millions of years.) Here, a base two with exponents were used to half units and identify more refined points.

What we consistently found was evidence for increasingly complexity, an acceleration of events logarithmically, such that Major Change kept happening in less time, or more precisely half the time as the previous change. That means that change occurred in pulsed logarithmic intervals.

I have attempted to why this happens, but that is better left for another work. For now, we leave this task largely to readers to wonder about and make their own theories. We will however take on the humble and diligent task of assembling the evidence and ordering it on a logarithmic scale. Such as a scale can be updated, and I have begun some of that process. What I found was that new evidence surfacing was increasingly confirming of the logarithmic structuring outlined here.

We will let you judge yourself as to how sufficient the evidence is and what that might say about the universe. It is an exciting process, and one which can only be "in process," to be continually updated and refined by others such as yourself. Let us then discover together.

Read more at https://www.allroneofus.com/.

Also by Michael A. Susko

A Couple Through Time

Down Below & the Archon's Castle

Up Above & the Runaway

Across the Gulf & Journey Into Un-Time

On the Bay & a Child Found

Down New River & Another World

In the Wild & Do One Wild Thing

On the Mountain & Two Are Missing

To the Beginning & Journey Through Here

Archetypal Worlds

Giant Under the Mountain & the Mystery of Sacrifice

The Alien's Gift

The Gold People

Spider Woman and the Timeroc

Darkwood and Dual with the Shadow Side

Quill Ears & the Other Earth

Alwon in Another World: An Archetypal Voyage

Line In the Wall

Beyond and a New World

Beyond & a New World: I Cosmos

Beyond & A New World: II Bios

Beyond & a New World: III Mythos

Beyond & A New World: IV Logos

Biographic Book of Tens

Ten Discoveries from Biology to Spirituality: Hidden & Life-Giving Connections

Ten Times We Almost Died

Ten Sayings to Guide Our Lives

Ten Mystery Photos: Personal & Cosmological Reflections

Ten Mementos on Our Desk: Remembering Moments

Ten Metadiscoveries We Have Made

Four Years

A Survival Manual for the Next Four Years

What—Me Worry? The Next Four Years

Flirting with the Fox: The Next Four Years

New Holidays are Coming! The Next Four Years

New Monuments are Coming! The Next Four Years

New Bible Stories are Coming! The Next Four Years

New Fairy Tales Are Coming! The Next Four Years

Why I Don't Need to Repent: Four More Years!

Haikus and Photos

Flowers and Haikus
Haikus and Photos: Guatemalan Highlands
Haikus and Photos: Water Birds and Reflections
Haikus and Photos: Seasons of New River
Haikus and Photos: Yosemite Wilderness
Haikus and Photos: California Coast
Haikus and Photos: Canadian Rockies
Haikus and Photos: Hawaii's Exotic Landscapes
Haikus and Photos: Vienna: People, Buildings and Art
Haikus and Photos: Slovakian Castles and Hamlets
Haikus and Photos: Berlin, Light and Dark
Haikus and Photos: New Orleans, City of Immigrants
Haikus and Photos: Antietam Wind and Spirits
Haikus & Photos: Plant Abstractions
Haikus and Photos: Appalachian Beauty
Haikus and Photos: Urban Farm in Sandtown
Haikus and Photos: New York Heights and Ground
Haikus & Photos: Santa Fe Fractal-Pueblo Spirtuality
Haikus and Photos: Monticello's Double Vision

Little Lion

The Lion and the Chameleon
The Elephant and the Chameleons
Little Lion and His Friends
Little Lion In Flower World
Gurr in Arid Land
Little Dragon Goes to China

Logarithmic Evolutionary Time

The Logarithmic Unfolding of Astronomical Time: From Quantum Reality to Life
Ten Pulses of Evolution & the Surprising Nature of Evolutionary Time
Ten Nodes of Symbolizing Humanity: From the First Hominin to the Neolithic

Mystery Stone Explorations
What Does ChatGPT See in the Mystery Stone?

Mystery Stone Formation
Mystery Formation at Penn Bluff: Resonance with Eastern Woodland Cosmology

Nature Haikus & Photos
Haikus and Photos: Butterflies and Flowers
Haikus and Photos: A Cardinal's Life
Haikus and Photos: Two Racoons at Play
Haikus and Photos: Hawaiian Green Sea Turtle
Haikus and Photos: Irises in the Rain
Haikus and Photos: The Mystic Iris
Haikus and Photos: Tulips Opening

Paired Series of A Couple Through Time
Down Below & Up Above
Across the Gulf & On the Bay

Down New River and In the Wild
On the Mountain & To the Beginning

Rosetta Key

A Rosetta Key for History: Generations Revealing the Phases of Time
A Rosetta Key For U.S. History: Renewal or Rigidification?
A Rosetta Key for Ancestral Pueblo History: Phases of an Indigenous Civilization

Second Mystery Stone from the Shenandoah

Haikus & Photos: 2nd Shenandoan Mystery Stone
Haikus and Photos: 2nd Mystery Stone 3-D Forms

Servants of the Flower World

Flower Amid the Devastation: Servants of the Flower World, Book I
Questers in Flower World: Servants of the Flower World, Book II
Fight for This World: Servants of the Flower World, Book III
Journey to the Underworld & The Circle of Emptiness
The Edge of Return

Shenandoan Stone Explorations

Mystery Stone from the Shenandoah: Analyzed with Eastern Woodland Cosmology
Philosopher Stone from the Lower Shenandoah
Beyond the Portal: From Within the Mystery Stone
The Mystery of Essence

Gifts from the Indigenous: Another Way of Viewing & Acting in the World

Shenandoan Stone: Haikus & Photos

Haikus and Photos: Cosmogram from the Shenandoah

Haikus and Photos: Woodland Mystery Stone and World Archetypes

Haikus and Photos: Skeletal Human and Mississippian Art

Haikus and Photos: Mystery Stone's Animal Forms

Haikus and Photos: Plant Forms and Mystery Stone

Stone Formation at Penn Bluff

Haikus and Photos: Presence at Penn Bluff

Haikus & Photos: Mystery Forms at Penn Bluff

Haikus and Photos: Essences at Penn Bluff

Haikus and Photos: World Archetypes at Penn Bluff

The Dreaming Series

Sleek Back & Salamander Dreaming

Streak & Cave Bear Dreaming

Moby & Marsupial Mole Dreaming

The Dream World Trilogy

Delphi, the Time Thief, and the Dream World

Detinna and the Cave God

The Resistance & the Empire

Transformational Stories

Caseness and Narrative: Contrasting Approaches to People Psychiatrically Labeled

Transformative Experiences, Psychiatric Research, and Informed Consent

Transformational Stories: Voices for True Healing in Mental Health

Worlds to the Side

A Couple Through Time, I. From Down Below to On the Bay

A Couple Through Time, II From Down New River to Journey Through Here

Writings from Street People

Street Images

Street Images II

Standalone

Little People & the Curious Time-Riding Animal

Animal Spell: A Journey Through Darkness & Light

Child of the Elements

The Firekeeper & The Spirit of the Long Night

Life's Dynamic Vulnerability, and Its Natural Extravagance

Alien Ally

The Generation of Life: Imagery, Ritual and Experiences in Deep Caves

Twelve Suspects: And the Case of the Missing Body

2084: Clash of Cults
Bats in the Future
Guard of the Dead & Servant of the Living
The Imagination Being
Ten Traits of Empire That Every Person Should Know
Aging and Renewal: Actualizing Our LIves
The Meaning, Beauty & Mystery of Dreams: Seven Guidelines and Seven Tools for Listening
The Fragility of Evolution: Re-envisioning Life's Creative Processes
Ways We May Be Surprised by Heaven
Why Go Slow When You Can Hurl to Your Destruction?
Stages of the Human Life Cycle: A Novel Logarithmic Perspective
Fifteen Amazing Things About the Body
The Human Body as Evolutionary Treasures
What does ChatGPT see in the Mystery Photos?
Agents of Immortality: Six Ways To Living Forever
The Logarithmic Nature of Time: From the Start of Life to Symbolizing Humanity
Animal Dreaming: The Adventures of Finn
Chameleon Adventures
Dream World Trilogy
As Lover, Lion, & Phoenix
Generational Key to History: Tracing Phases From Ancient Egypt to America
Planetary Revolution: The Next Stage in Human Awareness!
Down Below to the Beginning
New Ways Are Coming! The Next Four Years
Seven Ways to Prove You're a Christian Leader
From Sunfish to Son: The Gifts of Our Lives
Cosmos & Bios: Beyond & a New World
Astronomical & Biological Unfolding of Time
Servants of the Flower World: Book I-III
Mystery Photos: Human & AI Reflections

Down Below to a New World: A Couple Through Time
Human & AI Reflections on a Mystery Stone
The Unfolding of Time: A Logarithmic Model for Astronomical, Biological, and Cultural Evolution
Servants of the Flower World: The Complete Saga I-V
Servants of the Flower World, Book IV-V: Journey to the Underworld & The Edge of Return
Generational Mapping Through Phases: Bridging Ancient Civilizations, U.S. and Ancestral Pueblo History
Moments of Sacrificial Love That the World May Live: Tales from Servants of the Flower World
Cosmos, Bios, Mythos & Logos: An Unfolding of Time Saga—in Beyond and a New World
Mythos & Symbolic Humanity: Paired Literary–Scientific Visions — Language, Stone-Shaping, and the Birth of Culture
Cosmos & Astronomical Pulses in Time: Paired Literary–Scientific Visions — Galaxies, Planets & Life
Bios & Ten Pulses of Evolution: Paired Literary–Scientific Visions — From the Origin of Life to the First Hominids
Story & Stone I: Paired Literary & Anthropological Visions: Servants of the Flower World (Books I-II) & Mystery Stone
Story & Stone II: Paired Literary & Anthropological Visions: Servants of the Flower World (Books III-V) & Ancestral Pueblo History
Epic & Opus — Paired Visions of Time: Cosmos, Bios, Mythos
A Rosetta Key for Neolithic History: Birth of Phasing—Triadic Generations at Çatalhöyük
The Birth and Unfolding of Phases in History: From Neolithic to World, Pueblo, and U.S. History
Astronomical Pulses to History's Phases: The Unfolding & Expansion of Time
A Rosetta Key for The Kingdom of Mukish: The Crossroads Civilization that Bloomed

Paired Epics, Life and Afterlife of a Couple Through Time

Watch for more at https://www.allroneofus.com/.

About the Author

The author, having degrees in philosophy and psychology, has taught courses in history, Indigenous studies and dream interpretation. He has helped found and taught in a progressive charter school which used arts integration and emphasized the sciences. In his own research, he has identified and published logarithmic patterning of time in biological and human cultural evolution. Now, with the finding of new discoveries, he hopes to extend his work to the history of the astronomical universe.

Read more at https://www.allroneofus.com/.

AllrOneofUs Publishing

About the Publisher

AllRoneofUs Publishing seeks out work that offers a novel and qualitative contribution to world literature—the kind of work that may endure across generations.

We are especially drawn to voices that are often underheard, overlooked, or nearly lost—including authors in the later chapters of life who have made exemplary contributions that remain unrecognized. Our imprint serves as a sanctuary for works of insight, humanity, and lasting value.

Among those we have been honored to publish, we recommend:*Rich Mullin's Ethics and the Full-Breasted Richness of Life*,*John Susko's Flowers of the Night: Musings from a Sentimental Son*,and *Dr. Curtis Adams' Psychosis and the Humpty Dumpty Story*.

www.ingramcontent.com/pod-product-compliance
Lightning Source LLC
LaVergne TN
LVHW022109200926
848481LV00003B/1116
* 9 7 9 8 2 2 4 8 7 6 6 3 1 *